INTEGRITY
IN SCIENTIFIC RESEARCH

Creating an Environment That Promotes

Responsible Conduct

Committee on Assessing Integrity in Research Environments

Board on Health Sciences Policy
and
Division of Earth and Life Studies

INSTITUTE OF MEDICINE
NATIONAL RESEARCH COUNCIL
OF THE NATIONAL ACADEMIES

THE NATIONAL ACADEMIES PRESS
Washington, D.C.
www.nap.edu

THE NATIONAL ACADEMIES PRESS • 500 Fifth Street, N.W. • Washington, DC 20001

NOTICE: The project that is the subject of this report was approved by the Governing Board of the National Research Council, whose members are drawn from the councils of the National Academy of Sciences, the National Academy of Engineering, and the Institute of Medicine. The members of the committee responsible for the report were chosen for their special competences and with regard for appropriate balance.

Support for this project was provided by the Office of Research Integrity, U.S. Department of Health and Human Services. The views presented in this report are those of the Institute of Medicine and National Research Council Committee on Assessing Integrity in Research Environments and are not necessarily those of the funding agencies.

International Standard Book Number: 0-309-08523-3; 0-309-08479-2 (pbk.)

Library of Congress Control Number: 20021102-17

Additional copies of this report are available for sale from the National Academies Press, 500 Fifth Street, N.W., Box 285, Washington, DC 20055; (800) 624-6242 or (202) 334-3313 (in the Washington metropolitan area); Internet, http://www.nap.edu.

For more information about the Institute of Medicine, visit the IOM home page at: **www.iom.edu.**

THE NATIONAL ACADEMIES
Advisers to the Nation on Science, Engineering, and Medicine

The **National Academy of Sciences** is a private, nonprofit, self-perpetuating society of distinguished scholars engaged in scientific and engineering research, dedicated to the furtherance of science and technology and to their use for the general welfare. Upon the authority of the charter granted to it by the Congress in 1863, the Academy has a mandate that requires it to advise the federal government on scientific and technical matters. Dr. Bruce M. Alberts is president of the National Academy of Sciences.

The **National Academy of Engineering** was established in 1964, under the charter of the National Academy of Sciences, as a parallel organization of outstanding engineers. It is autonomous in its administration and in the selection of its members, sharing with the National Academy of Sciences the responsibility for advising the federal government. The National Academy of Engineering also sponsors engineering programs aimed at meeting national needs, encourages education and research, and recognizes the superior achievements of engineers. Dr. Wm. A. Wulf is president of the National Academy of Engineering.

The **Institute of Medicine** was established in 1970 by the National Academy of Sciences to secure the services of eminent members of appropriate professions in the examination of policy matters pertaining to the health of the public. The Institute acts under the responsibility given to the National Academy of Sciences by its congressional charter to be an adviser to the federal government and, upon its own initiative, to identify issues of medical care, research, and education. Dr. Harvey V. Fineberg is president of the Institute of Medicine.

The **National Research Council** was organized by the National Academy of Sciences in 1916 to associate the broad community of science and technology with the Academy's purposes of furthering knowledge and advising the federal government. Functioning in accordance with general policies determined by the Academy, the Council has become the principal operating agency of both the National Academy of Sciences and the National Academy of Engineering in providing services to the government, the public, and the scientific and engineering communities. The Council is administered jointly by both Academies and the Institute of Medicine. Dr. Bruce M. Alberts and Dr. Wm. A. Wulf are chair and vice chair, respectively, of the National Research Council.

www.national-academies.org

Project Staff

THERESA M. WIZEMANN, Study Director, Board on Health Sciences Policy

MEHREEN N. BUTT, Senior Project Assistant, Board on Health Sciences Policy

FREDERICK J. MANNING, Senior Program Officer, Board on Health Sciences Policy

ROSEMARY CHALK, Senior Program Officer, Board on Health Care Services

Auxiliary Staff

ANDREW POPE, Director, Board on Health Sciences Policy

DALIA GILBERT, Research Assistant, Board on Health Sciences Policy

ALDEN CHANG, Administrative Assistant, Board on Health Sciences Policy

CARLOS GABRIEL, Financial Associate

ROBIN SCHOEN, Program Officer, Board on Life Sciences, Division on Earth and Life Sciences

Consulting Writer

KATHI E. HANNA

Copy Editors

TOM BURROUGHS
MICHAEL K. HAYES

Reviewers

This report has been reviewed in draft form by individuals chosen for their diverse perspectives and technical expertise, in accordance with procedures approved by the National Research Council's (NRC's) Report Review Committee. The purpose of this independent review is to provide candid and critical comments that will assist the institution in making its published report as sound as possible and to ensure that the report meets institutional standards for objectivity, evidence, and responsiveness to the study charge. The review comments and draft manuscript remain confidential to protect the integrity of the deliberative process. We wish to thank the following individuals for their review of this report:

JOHN F. AHEARNE, Sigma Xi Center, The Scientific Research Society
PAUL J. FRIEDMAN, University of California, San Diego
C. KRISTINA GUNSALUS, University of Illinois at Urbana-
    Champaign,
RUSSEL E. KAUFFMAN, The Wistar Institute
DAVID KORN, Association of American Medical Colleges
JEFFREY D. KOVAC, University of Tennessee
MARCEL C. LAFOLLETTE, George Washington University
MARY FAITH MARSHALL, Kansas University Medical Center
RICK ANTONIO MARTINEZ, Johnson and Johnson
JUDITH P. SWAZEY, The Acadia Institute

Although the reviewers listed above have provided many constructive comments and suggestions, they were not asked to endorse the conclusions or recommendations, not did they see the final draft of the report before its release. The review of this report was overseen by **BERNARD LO**, University of California, San Francisco, appointed by the Institute of Medicine, and **HAROLD C. SOX**, Annals of Internal Medicine, appointed by the NRC's Report Review Committee, who were responsible for making certain that an independent examination of this report was carried out in accordance with institutional procedures and that all review comments were carefully considered. Responsibility for the final content of this report rests entirely with the authoring committee and the institution.

Acknowledgments

The committee is indebted to the researchers and administrators who presented informative talks to the committee and participated in lively discussions at the open meetings, including Melissa Anderson, Stephanie Bird, Ruth Fischbach, Peter Fiske, Barbara Mishkin, Howard Schachman, Joan Schwartz, Harold Varmus, Bart Victor, and Peter Yeager (see Appendix A for affiliations and discussion topics). The committee is grateful to Barbara Brittingham, Steven Crow, Beth Fischer, Alasdair MacIntyre, Jean Morse, George Peterson, James Rogers, David Smith, David Stevens, and Naomi Zigmond, who graciously made themselves available by phone and e-mail for consultation and technical advice, and to Kenneth Pimple and David Guston, who were commissioned to prepare technical literature reviews and historical reviews (see Appendix A). Thanks to Diane Waryold and the Center for Academic Integrity for kindly providing their Academic Integrity Assessment Guide. The committee also thanks the Institute of Medicine and National Research Council staff who presented overviews of previous Academy work on integrity in research, including Rosemary Chalk, Robin Schoen, and Debbie Stine.

Many thanks to Bruce Alberts, president of the National Academy of Sciences and chair of the National Research Council, Kenneth Shine, then president of the Institute of Medicine, Clyde Behney, deputy director of the Institute of Medicine, and Andrew Pope, director of the Institute of Medicine Board on Health Sciences Policy, for advice and guidance in addressing the task. Thanks also to Kathi Hanna, Michael Hayes, and Tom Burroughs for assistance with editing the text of the report.

The committee also wishes to thank Nicholas Steneck, University of Michigan, for his contributions during the early stages of the study, and Jennifer Rietfors, an intern at the American Association for the Advancement of Science, for assistance with information on accrediting bodies.

This report was made possible by the generous support of the Office of Research Integrity, U.S. Department of Health and Human Services. Thanks to Chris Pascal and Larry Rhoades for providing background information, advice, and encouragement throughout the course of the study.

Contents

APPENDIXES

Tables, Figures, and Boxes

TABLES

FIGURES

BOXES

Executive Summary

The pursuit and diffusion of knowledge enjoy a place of distinction in American culture, and the public expects to reap considerable benefit from the creative and innovative contributions of scientists. Most Americans have a positive attitude toward science and technology and are willing to demonstrate their support through public investments in science and research institutions. Public funding is based on the principle that the public good is advanced by science conducted in the interest of humanity. Such support is qualified, however. The public will support science only if it can trust the scientists and the institutions that conduct research. Major social institutions, including research institutions, are expected to be accountable to the public. Fostering an environment that promotes integrity in the conduct of research is an important part of that accountability. As a consequence, it is more important than ever that individual scientists and their institutions periodically assess the values and professional practices that guide their research as well as their efforts to perform their work with integrity.

Considerable effort has been devoted to the task of defining research misconduct and elaborating methods for investigating allegations of misconduct. Much less attention has been devoted, however, to the task of fostering a research environment that promotes integrity. This report focuses on the research environment and attempts to define and describe those elements that enable and encourage unique individuals, regardless of their role in the research organization or their backgrounds on entry, to act with integrity. Although integrity and misconduct are related, the focus of this report is on integrity. The Institute of Medicine (IOM) Committee on Assessing Integrity in Research Environments, which prepared this report, does not discuss or draw conclusions about current or proposed regulations or definitions relating to misconduct. The committee's goal was to define the desired outcomes and set forth a set of initiatives that it believes will enhance integrity in the research environment. The committee considered approaches that can be used to promote integrity and methods that can be used to assess the effectiveness of those efforts. The majority of these approaches and methods can and should be initiated as soon as feasible and administered by research institutions themselves so that government regulation will not be required.

CHARGE TO THE COMMITTEE

In January 2001, IOM, in collaboration with the National Research Council's Division on Earth and Life Studies, formed the Committee on Assessing Integrity in Research Environments, in response to a request from the Office of Research Integrity (ORI) of the U.S. Department of Health and Human Service (DHHS). In general, the committee was charged with addressing the need of DHHS to track the state of integrity in the research environment. More specifically, the committee was asked to do the following:

1. define the concept "research integrity";
2. describe and define the concept "research environment";
3. identify elements of the research environment that promote research integrity;
4. indicate how the elements may be measured;
5. suggest appropriate methodology for collecting the data;
6. cite appropriate outcome measures;
7. make recommendations regarding the adoption and implementation by research institutions, government agencies, scientific societies, and others (as appropriate) of those elements of the research environment identified to promote integrity in research; and
8. convene a public meeting to discuss the IOM report, its recommendations, and potential strategies for their implementation.

To respond to the charge, the committee explored various data sources in its effort to provide ORI with a means for tracking the state of integrity in the research environment. In addition to reviewing the professional literature, the committee also reviewed numerous reports, regulations, and guidelines of the federal government and articles and editorials in the popular press. The committee invited experts to make public presentations, commissioned background papers, and sought additional technical assistance from knowledgeable individuals.

OVERARCHING CONCLUSIONS

Several overarching conclusions emerged as the committee addressed DHHS's need to develop means for assessing and tracking the state of integrity in the research environment:

- Attention to issues of integrity in scientific research is very important to the public, scientists, the institutions in which they work, and the scientific enterprise itself.
- No established measures for assessing integrity in the research environment exist.
- Promulgation of and adherence to policies and procedures are necessary, but they are not sufficient means to ensure the responsible conduct of research.
- There is a lack of evidence to definitively support any one way to approach the problem of promoting and evaluating research integrity.
- Education in the responsible conduct of research is critical, but if not done appropriately and in a creative way, then education is likely to be of only modest help and may be ineffective.
- Institutional self-assessment is one promising approach to assessing and continually improving integrity in research.

The committee found that existing data are insufficient to enable it to draw definitive conclusions as to which elements of the research environment promote integrity. The elements discussed in the report appear to be associated with integrity in research, but the specific contribution of each element remains poorly defined. Empirical studies evaluating the ethical climate before and after implementation of specific policies or practices are lacking.

Because of the limited empirical data on factors influencing responsible conduct in the scientific environment, the committee drew on more general theory (e.g., theories of organizational behavior, ethical decision making, and adult learning) to formulate the suggestions presented in the report. The findings and conclusions are based on the committee's collective knowledge and experience after its review of the literature in the science and business arenas as well as its discussions with experts who presented talks at the committee's open meetings.

FINDINGS AND RECOMMENDATIONS

Integrity in Research

Integrity in research is essential for maintaining scientific excellence and for keeping the public's trust. Integrity characterizes both individual researchers and the institutions in which they work. The concept of integrity in research cannot, however, be reduced to a one-line definition. For a scientist, integrity embodies above all the individual's commitment to intellectual honesty and personal responsibility. It is an aspect of moral character and experience. For an institution, it is a commitment to creating an environment that promotes responsible conduct by embracing standards of excellence, trustworthiness, and lawfulness and then assessing whether researchers and administrators perceive that an environment with high levels of integrity has been created. Many practices are likely to promote responsible conduct (see Box 1). Individuals and institutions should use these practices with the goal of fostering a culture in which high ethical standards are the norm, ongoing professional development is encouraged, and public confidence in the scientific enterprise is preserved.

The Research Environment

The research environment changes continually, and these changes influence the culture and conduct of research. As with any system being scientifically examined, the research environment itself contains variables and constants. The most unpredictable and influential variable is the individual scientist. The human contribution to the research environment is greatly shaped by each individual's professional integrity, which in turn is influenced by that individual's educational background and cultural and ethical upbringing and the resulting values and attitudes that contribute to identity formation, unique personality traits, and ethical decision-making abilities.

Since each individual researcher brings unique qualities to the research environment, the constants must come from the environment itself. Research institutions should consistently and effectively provide training and education, policies and procedures, and tools and support systems. Institutional expectations should be unambiguous, and the consequences of one's conduct should be clear. Institutional leaders should set the tone for the institutions with their own actions. Those in leadership positions should explicitly and actively endorse, and participate in, activities that foster responsible conduct of research. Anyone needing assistance should have ready access to knowledgeable leaders and should be able to seek help and advice without fear of retribution. Institutions re-

---

BOX 1
Integrity in Research

Individual Level

For the individual scientist, integrity embodies above all a commitment to intellectual honesty and personal responsibility for one's actions and to a range of practices that characterize the responsible conduct of research, including

- intellectual honesty in proposing, performing, and reporting research;
- accuracy in representing contributions to research proposals and reports;
- fairness in peer review;
- collegiality in scientific interactions, including communications and sharing of resources;
- transparency in conflicts of interest or potential conflicts of interest;
- protection of human subjects in the conduct of research;
- humane care of animals in the conduct of research; and
- adherence to the mutual responsibilities between investigators and their research teams.

Institutional Level

Institutions seeking to create an environment that promotes responsible conduct by individual scientists and that fosters integrity must establish and continuously monitor structures, processes, policies, and procedures that

- provide leadership in support of responsible conduct of research;
- encourage respect for everyone involved in the research enterprise;
- promote productive interactions between trainees and mentors;
- advocate adherence to the rules regarding all aspects of the conduct of research, especially research involving human participants and animals;
- anticipate, reveal, and manage individual and institutional conflicts of interest;
- arrange timely and thorough inquiries and investigations of allegations of scientific misconduct and apply appropriate administrative sanctions;
- offer educational opportunities pertaining to integrity in the conduct of research; and
- monitor and evaluate the institutional environment supporting integrity in the conduct of research and use this knowledge for continuous quality improvement.

---

quire support mechanisms, such as ombudspersons, that research team members can turn to with concerns about integrity, including reporting suspected misconduct.

The committee found no comprehensive body of research or writing that can guide the development of hypotheses regarding the relationships between the research environment and the responsible conduct of research. Thus, the committee drew on more general theoretical and research literature to inform its discussion. Relevant literature was found in the areas of organizational behavior and processes, ethical cultures and climates, moral development, theories of adult learning and educational practices, and professional socialization.

Viewing the research environment as an open-systems model,[1] which is often used in general organizational and administrative theory, enables one to hypothesize how various components affect integrity in research (Figure 1). Inputs of funds and other resources can influence behavior both positively and negatively. The organizational structure and processes that typify the mission and activities of the organization can either promote or detract from the responsible conduct of research. The culture and climate that are unique to an organization both promote and perpetuate certain behaviors. Finally, the external environment (Figure 2), over which individuals and often institutions have little control, can affect behavior and alter institutional integrity for better or for worse.

Fostering Integrity

Institutions should develop a multifaceted approach to promoting integrity in research appropriate to their research environments. At present, research organizations rely on a variety of methods. They establish organizational components to comply with regulations imposed by an external environment; they offer educational programs to teach the elements of the responsible conduct of research; and they implement policies and procedures that delineate the normative practices of responsible conduct of research and establish their criteria for rewards and recognition. In addition, organizations engage in activities that help establish an internal climate and organizational culture that are either supportive of or ambivalent toward the responsible conduct of research. Of course, these various approaches are not mutually exclusive, nor should they be. A number of programs and activities, integrated across organizational levels, should be in place in order to maximize the impact on the research environment and support the responsible conduct of research.

---

[1]The open-systems model depicts the various elements of a social organization, including the external environment, the organizational divisions or departments, the individuals comprising those divisions, and the reciprocal influences between the various organizational elements and the external environment (see Chapter 3).

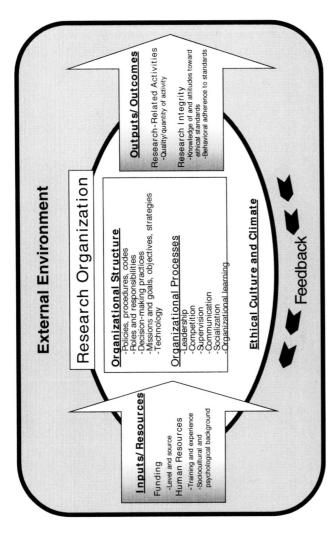

FIGURE 1 Open-systems model of the research organization. This model depicts the internal environmental elements of a research organization (white oval), showing the relationships among the inputs that provide resources for organizational functions, the structures and processes that define an organization's operation, and the outputs and outcomes of an organization's activities that are carried out by individual scientists, research groups or teams, and other research-related programs. All of these elements function within the context of an organization's culture and climate. The internal environment is affected by the external environment (shaded area; see Figure 2 for further detail). The system is dynamic, and, as indicated by the feedback arrow, outputs and outcomes affect future inputs and resources.

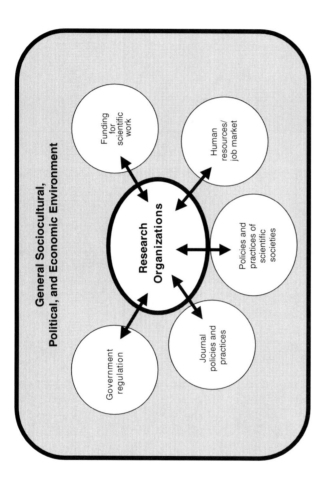

FIGURE 2 Environmental influences on integrity in research that are external to research organizations. The *external-task environment* includes all of the organizations and conditions that are directly related to an organization's main operations and technologies. The double arrows depict the interrelatedness between the research organization and the various external influences (unshaded circles) that are hypothesized to have an impact on integrity in research. The *general environment* has a more indirect impact on an organization. The systems and subsystems of the external-task environment are embedded within the larger, general sociocultural, political, and economic environment (shaded area). Although not specifically shown in this figure, it is important to recognize that relationships exist between and among the elements within the external environment.

Promoting Integrity in Research through Education

The provision of instruction in the responsible conduct of research need not be driven by federal mandates, for it derives from a premise fundamental to doing science: the responsible conduct of research is not distinct from research; on the contrary, competency in research *encompasses* the responsible conduct of that research and the capacity for ethical decision making. For lasting change in ethical climate to occur, changes in curriculum content alone are not sufficient. Attention also needs to focus on how education in the responsible conduct of research is conducted. Indeed, integrity in research should be developed within the context of other aspects of an overall research education program. The committee believes that doing so will be the best way to accomplish the following five objectives for graduate students and postdoctoral fellows:

1. emphasize responsible conduct as central to conducting good science;
2. maximize the likelihood that education in the responsible conduct of research influences individuals and institutions rather than merely satisfies an item on a checkoff list necessary for that institution;
3. impart essential rules and guidelines regarding responsible conduct of research in one's discipline and profession in context;
4. enable participants in the educational programs to develop abilities that will help them to effectively manage concerns related to responsible conduct of research that cannot be anticipated but that are certain to arise in the future; and
5. verify that the first four objectives have been met.

Teaching of the responsible conduct of research presents a special challenge because it requires a synthesis of ethics and science. When scientists and ethicists collaborate in the design and implementation of learning experiences, students come to appreciate the complexity of problems that arise in the practice of science. Furthermore, when instruction requires the application of norms (and the ethical theories that support them), values, and rules and regulations to the practical problems that arise in the day-to-day practice of science, learning is more likely to last and to transfer to new situations. It follows, then, that instruction in the responsible conduct of research by a team of faculty—or by a faculty member with expertise in both ethics and science—is optimal. When faculty take time from their scholarly work to provide practical instruction that draws on expertise from related fields, they demonstrate the importance of this educational task and its relevance to the practice of science. Research advisers play a central role in the education of their trainees in the responsible conduct of research, not only by what they teach, but also by their own conduct. The impact of educational efforts may be weakened if what is taught is not actively practiced by supervisors and administrators.

It should be noted that this report emphasizes the education of students and postdoctoral fellows, not because the committee believes that this is where a problem exists but, rather, because this is where the future lies. Thus, the model for providing instruction in the responsible conduct of research is taken from traditional programs for teaching students what is necessary for their performance as researchers: (1) start as soon as the students arrive; (2) make the instruction in this area part of everything they do, placing the education in the context of the research instead of making it a separate entity; (3) move from the simple to the complex; and (4) assess student competency. In this way, there is no mistaking the message: communicating well, obtaining employment and research grants, excelling in teaching and mentoring, engaging in ethical decision making, and behaving responsibly are at the core of being a researcher, in addition to sophisticated use of knowledge to plan and execute research.

Evaluation by Self-Assessment

To optimize the institutional approach to fostering the responsible conduct of research, it is critical that organizations simultaneously implement processes for evaluating their efforts, thereby establishing a basis for organizational learning and continuous quality improvement. Evaluation can be approached in a variety of ways. One way is to rely on external evaluators to determine compliance with regulatory controls. Another is to rely on a system of performance-based assessments that are initiated and implemented internally. Such assessments can also be used to meet the accountability requirements of outside funding and government sources. In addition, peer reviewers may be used in institutional self-assessment processes; assessments done by peer reviewers may or may not be associated with accreditation by external organizations.

Although the regulatory approach has led to some successes, the committee felt that a regulatory approach to fostering integrity in research has some important limitations. Such an approach increases the bureaucratization of science and requires documentation that institutions may find burdensome. Regulations often emphasize the areas of common agreement and reduce important concerns to rules and procedures, rather than foster a deep understanding of the ethical issues involved and the variety of sophisticated approaches available to address those issues. The adoption of new regulations and the creation of institutional and governmental oversight offices increase the cost of doing science and add to the administrative costs of research centers without necessarily creating a commensurate benefit. In addition, once regulations are adopted, it is difficult to change them. Regulatory frameworks reduce the flexibility of institutions and individuals to respond to research opportunities.

At this time, neither the executive nor legislative branches of government has established a regulatory framework to foster integrity in scientific research, and the committee does not believe such a framework would be desirable or comparable to the system that has been put into place to address misconduct in science or the use of institutional review boards. Rather, the committee endorses the principle of self-assessment as a component of formal performance appraisals of academic departments and of individual faculty members.

The committee proposes that research institutions work with established accrediting bodies to develop mechanisms for incorporating institutional self-study for integrity in research into the overall accreditation process. The processes of established accrediting bodies are expected to be more effective and more cost-efficient than those of a new entity, whose establishment would be seen as one more administrative burden, and thus would encourage cynicism.

If institutional cultures are to be changed, then both the call for change and its implementation must come from research institutions. An important next step will be for universities and university associations, working together, to acknowledge the importance of conducting research and research education in an environment of high integrity and developing an evaluative process based on self-study.

Methods and Measures

Gaining the methodological expertise needed to carry out research on the relationship between the research environment and integrity in research will require the development and validation of measures, particularly indicators that are observable and quantifiable within the research environment. For example, existing means of conceptualizing and measuring the organizational climate will have to be adapted to this specific context of the assessment of the ethical climate within the research environment.

Furthermore, to measure the effectiveness of efforts related to fostering integrity in the research environment, specific outcomes must be identified and defined within this context. Next, either new instruments must be designed and validated, or existing outcome measures must be modified and validated for the assessment of the ethical climate within the research environment. This development of reliable and valid measures will take considerable time and effort, but it is a necessary step in a re-

BOX 2
Recommendations

Future Research

RECOMMENDATION 1: Funding agencies should establish research grant programs to identify, measure, and assess those factors that influence integrity in research.

• The Office of Research Integrity should broaden its current support for research to fund studies that explore new approaches to monitoring and evaluating the integrity of the research environment.

• Federal agencies and foundations that fund extramural research should include in their funding portfolios support for research designed to assess the factors that promote integrity in research across different disciplines and institutions.

• Federal agencies and foundations should fund research designed to assess the relationship between various elements of the research environment and integrity in research, including similarities and differences across disciplines and institutions.

Institutional Commitment to Integrity

RECOMMENDATION 2: Each research institution should develop and implement a comprehensive program designed to promote integrity in research, using multiple approaches adapted to the specific environments within each institution.

• It is incumbent upon institutions to take a more active role in the development and maintenance of climate and culture within their research environments that promote and support the responsible conduct of research.

• The factors within the research environment that institutions should consider in the development and maintenance of such a culture and climate include, but are not limited to, supportive leadership, appropriate policies and procedures, effective educational programs, and evaluation of any efforts devoted to fostering integrity in research.

• Federal research agencies and private foundations should work with educational institutions to develop funding mechanisms to provide support for programs that promote the responsible conduct of research.

Education

RECOMMENDATION 3: Institutions should implement effective educational programs that enhance the responsible conduct of research.

• Educational programs should be built around the development of abilities that give rise to the responsible conduct of research.

• The design of programs should be guided by basic principles of adult learning.

• Integrity in research should be developed within the context of other relevant aspects of an overall research education program, and instruction in the responsible conduct of research should be provided by faculty who are actively engaged in research related to that of the trainees.

Institutional Self-Assessment

RECOMMENDATION 4: Research institutions should evaluate and enhance the integrity of their research environments using a process of self-assessment and external peer review in an ongoing process that provides input for continuous quality improvement.

- The importance of external peer review of the institution cannot be overemphasized. Such a process will help to ensure the credibility of the review, provide suggestions for improvement of the process, and increase public confidence in the research enterprise.
- Effective self-assessment will require the development and validation of evaluation instruments and measures.
- Assessment of integrity and the factors associated with it (including educational efforts) should occur at all levels within the institution—for example, at the institutional level, the research unit level, and the individual level. At the individual level, assessment of integrity should be an integral part of regular performance appraisals.
- As with any new program, a phase-in or pilot testing period is to be expected, and the assessment and accreditation process should be continually modified as needed based on results of these early actions.

RECOMMENDATION 5: Institutional self-assessment of integrity in research should be part of existing accreditation processes whenever possible.

- Accreditation provides established procedures, including external peer review, that can be modified to incorporate assessments of efforts related to integrity in research within an institution.
- Entities that currently accredit educational programs at institutions where research is conducted would be the bodies to also review the process and the outcome data from the institution's self-assessment of its climate for promotion of integrity in research. These include the six regional organizations that accredit institutions of higher education in the United States, as well as organizations that accredit professional schools or professional educational programs.
- Federal research agencies and private foundations should support efforts to integrate self-assessment of the research environment into existing accreditation processes, and they also should fund research into the effectiveness of such efforts.

RECOMMENDATION 6: The Office of Research Integrity should establish and maintain a public database of institutions that are actively pursuing or employing institutional self-assessment and external peer-review of integrity in research.

- This database should initially include institutions that receive funding for, or are actively engaged in, the development and validation of self-assessment instruments.

search process leading to a better understanding of the relationship between the research environment and integrity in research. Note that two distinct types of measures should be considered: measures that assess the integrity of the institution with respect to the conduct of research and measures that assess aspects of the integrity of the individual.

Existing methods and measures, examples of which are described in Appendix B, provide models that could be adapted to evaluate the factors of culture and climate that promote integrity in research. Appendix B also provides examples of measures that have been used successfully to assess learning outcomes in professional ethics programs and that could be adapted to the research environment. On the basis of the available information, the committee describes practices that promote the responsible conduct of research and presents a model that captures the key components of the research environment and their interactivity. This is relatively new territory, however, that needs to be examined systematically with greater precision.

Focusing on the Future

Research institutions bear the primary burden of promoting and monitoring the responsible conduct of research. They must consistently provide members of research teams with the resources they need to conduct research responsibly. These resources include leadership and example, training and education, and policies and procedures, as well as tools and support systems. Institutional behavior should be exemplary. What is expected of individuals should be unambiguous, and the consequences of one's conduct should be clear. Anyone needing assistance should have ready access to knowledgeable leaders. Individuals should be able to seek assistance without fear of retribution. Research institutions, accrediting agencies, and public and private organizations that fund or otherwise support research should collaborate to establish and ensure the integrity of the scientific research enterprise. The collection of specific empirical data on integrity in scientific research is essential to help institutions determine the effectiveness of their efforts to foster a research climate that promotes integrity. Such data will also aid institutions in the development of better programs and policies in the future.

Government oversight of scientific research is important, but such oversight, often in the form of administrative rules, typically stipulates what cannot be done; it rarely prescribes optimal performance. In essence, government rules define the floor of expected behavior. More, however, should be expected from scientists when it comes to the responsible conduct of research. By appealing to the consciences of individual scientists, the scientific community as a whole should seek to evoke the highest possible standards of research behavior. When institutions committed to promoting integrity in research support those standards, the likelihood of creating an environment that promotes the responsible conduct of research is greatly enhanced. It is essential that institutions foster a culture of integrity in which students and trainees, as well as senior researchers and administrators, have an understanding of and commitment to integrity in research.

1

Introduction

"Rather fail with honor than succeed by fraud."

Sophocles

*"Most people say that it is the intellect which makes a great scientist.
They are wrong: it is character."*

Albert Einstein

Most Americans have a positive attitude toward science and technology (NSF, 2000b), and public confidence in consumer products is boosted by claims that they are "scientifically tested" or "scientifically proven." Such support is qualified, however. The public will support science only if it can trust the people and the institutions that conduct research. Major social institutions, including research institutions, are expected to be accountable to the public (Grinnell, 1999a; IOM, 2001; Yarborough and Sharp, 2002). Fostering an environment that promotes integrity in the conduct of research is an important part of that accountability.

Because of the complexity, variability, and nature of scientific inquiry, the concept of integrity in research can be elusive, and its value cannot be easily assessed or measured. From the outside looking in, science is a quest for truth about the natural world. In reality, scientific "truth" is always tentative; and the means for testing it involve repetition, disclosure, sharing of information, and competition. Scientists understand that "truth" and "fact" are based on the weight of scientific evidence. "Facts" hold only until they are successfully challenged by additional evidence, after which they may be modified or interpreted differently. Research usually proceeds from a mélange of hypotheses and results based on previous experiments and knowledge. New results may support the proposed hypothesis, but they can never prove a general hypothesis or theory.

In this progression toward the truth, researchers strive to be objective; but prior knowledge, opinion, and personal biases can influence the selection of hypotheses and study design, the conduct of the research, and interpretation of the results (Grinnell, 1992; Macrina, 2000). These preconceptions can inform and improve research, but the existence of such preconceptions can also cause investigators to stretch, and sometimes exceed, the limits of acceptable behavior. Thus, recognition of preconceptions, biases, and the need for integrity in the research process is essential for maintaining scientific excellence and the public's trust. Integrity in research embodies above all an individual's commitment to intellectual honesty and personal responsibility and an institution's commitment to creating an environment that promotes responsible conduct (see Chapter 2).

Nevertheless, even the best scientific intentions may produce unverifiable results because of flawed hypotheses, inadequate technology, the faulty execution of research, or the incorrect interpretation of results. In fact, errors, responses to errors, and validation of errors are important elements of the scientific process. In testifying to the Institute of Medicine (IOM) Committee on Assessing Integrity in Research Environments, which prepared this report, former National Institutes of Health (NIH) Director Harold Varmus said, "The notion of truth in science is difficult.... We don't know what the truth is. We are working our way toward the truth; and we are dependent on data, and data can be misleading." Consequently, even the most promising of experiments, conducted by seasoned researchers, will frequently fail. An important aspect of integrity in research is how one deals with error and with studies conducted erroneously. How mistakes are dealt with may have an important impact on the ethical climate of a research environment.

Biomedical research is often the focus of scrutiny because its findings can have important implications for health, it is highly regulated, and it receives substantial public funding. Moreover, serious errors or misconduct in biomedical research can lead to dire or even lethal consequences for research subjects. Media coverage of integrity in research usually focuses on clinical research catastrophes, egregious conflicts of interest, and overt research misconduct (e.g., the falsification or fabrication of data and plagiarism).

Even though issues related to the conduct of research in the areas of health and disease are foremost in the minds of many people, responsible conduct is vital for all areas of science. For example, research in the physical, chemical, and environmental sciences leads to the innovative and more effective use of natural resources; the identification of new energy sources; the structural and mechanical safety of bridges, buildings, and various modes of transportation; and methods for managing and reducing the waste generated from the actions of humans and the machines that they create. Technology and communications are also important in health and health care but they are just as important in commerce, services, and defense.

CHARGE TO THE COMMITTEE

In January 2001, in response to a request from the Office of Research Integrity (ORI) of the U.S. Department of Health and Human Services (DHHS), IOM, in collaboration with the National Research Council's Division on Earth and Life Studies, formed the Committee on Assessing Integrity in Research Environments. In general, the committee was charged with addressing the need of DHHS to track the state of integrity in the research environment. More specifically, the committee was asked to do the following:

1. define the concept "integrity in research";
2. describe and define the concept "research environment";
3. identify elements of the research environment that promote integrity in research;
4. indicate how the elements may be measured;
5. suggest an appropriate methodology for collecting the data;
6. cite appropriate outcome measures;
7. make recommendations regarding the adoption and implementation by research institutions, government agencies, scientific societies, and others (as appropriate) of those elements of the research environment identified to promote integrity in research; and
8. convene a public meeting to discuss the IOM report, its recommendations, and potential strategies for their implementation.

The committee membership included research and university administrators; educators; and active researchers from academic, industry, association, and private research settings. These individuals brought expertise from a broad array of fields, including biology, biomedical research, chemistry, clinical research, evaluation research and methodology, medicine, medical education, physics, public policy, science education, science ethics, and sociology.

ORIGINS OF THE STUDY

Several specific cases of alleged fraud or other scientific misconduct have been widely covered by the press in the past decade; and the federal government, the National Academies, and numerous scientific societies have made a considerable effort to develop a definition of research misconduct and guidelines for handling allegations of research misconduct. Such allegations remain relatively rare, however, and investigations of misconduct are not likely to be sensitive indicators of the changes in the research environment that might discourage misconduct and promote integrity in science.

In 1992, the National Academies' Committee on Science, Engineering, and Public Policy (COSEPUP) published *Responsible Science: Ensuring the Integrity of the Research Process* (NAS, 1992). That report was published after years of debate about policy over integrity in research that included serious allegations, congressional hearings, and media coverage.[1] That debate included controversy over the Office of Scientific Integrity (OSI), which DHHS created in 1989. OSI was immediately criticized for being overzealous in its investigations and inflexible in its process (Davis, 1991; Hamilton, 1991). A high-profile court case challenged the due process and legitimacy of OSI, and a later reorganization of the office consolidated activities into what is now the Office of Research Integrity (ORI) (Guston, 2000). As a result, new procedures provided a review and hearing process that gave accused individuals or institutions more opportunity to present their case (ORI, 1999).

In 1999, Secretary Shalala accepted the recommendations of the HHS Review Group on Research Misconduct and Research Integrity, and the primary focus of ORI shifted away from active investigation to education, oversight, and assurance, and, more recently, to research. The federal government firmly asserted its authority and discretion in setting conditions on the awarding of research grants. It now requires research institutions to have in place policies and procedures for handling allegations of misconduct, protecting whistle-blowers, and providing education in the responsible conduct of research for recipients of training grants (DHHS, 1995, pp. 21–24; DHHS, 2000).

At the same time that research institutions have augmented their ability to combat misconduct, the specific role of the federal government in investigating allegations has been legalized and limited. The federal government, particularly through the DHHS Departmental Appeals Board, has articulated clear standards for the adjudication of allegations of misconduct in research, with "misconduct" defined by the Office of Science and Technology Policy as "fabrication, falsification, or plagiarism in proposing, performing, or reviewing research, or in reporting research results" (OSTP, 2000, p. 76262). Institutions must conduct inquiries and investigations, and only when further fact finding is required by the federal government will the Office of the Inspector General at DHHS intervene.

---

[1]The committee asked David Guston, Rutgers University, to survey related events over the 10 years since publication of the COSEPUP report. Portions of this section were drawn from his report, which is summarized in Appendix C.

The Public Health Service (PHS) has also addressed research integrity. PHS regulation on responding to allegations of scientific misconduct states that "institutions shall foster a research environment that discourages misconduct in all research and that deals forthrightly with possible misconduct associated with research for which PHS funds have been provided or requested" (45 C.F.R. § 50.105, 2001). PHS has devoted considerable effort to the task of defining research misconduct and elaborating methods for investigation of allegations of misconduct. It has devoted much less attention to the task of fostering a research environment that promotes integrity. In 1999, the DHHS Review Group on Research Misconduct and Research Integrity recommended that "the role, mission, and structure of ORI change to become one of preventing misconduct and promoting integrity in research principally through oversight, education, and review of institutional findings and recommendations" (DHHS, 2000, p. 30601).

To provide an empirical basis for this new mission, ORI plans to develop a longitudinal database that tracks the state of integrity in research environments. ORI, PHS, and the extramural research community could use this database to guide development of education, prevention, and research programs related to the responsible conduct of research and to evaluate the effectiveness of those programs. In the absence of such data, the federal government and the scientific community will continue to manage issues related to integrity in research in a relative vacuum and will primarily rely on speculation and the infrequent individual case reports that receive notoriety. This regulation by crisis can lead to expensive and inefficient solutions when the scope of the actual problem is unknown.

Other Efforts to Foster Integrity

This report is not the first effort that IOM has made to address and assess the best means for fostering integrity in the research environment. In 1989, IOM published *The Responsible Conduct of Research in the Health Sciences*. Since then, the National Academies has published a number of reports, essays, and guides that have recommended actions to promote responsible research practices, including guides on education and training, mentoring, and careers in science. Key among these are volumes 1 and 2 of *Responsible Science: Ensuring the Integrity of the Research Process* (NAS, 1992, 1993) and *On Being a Scientist* (NAS, 1989a, 1995).

The 1992 report *Responsible Science: Ensuring the Integrity of the Research Process* recommended "individual scientists in cooperation with officials of research institutions should accept formal responsibility for ensuring the integrity of the research process. They should foster an environment, a reward system, and a training process that encourage responsible research practices" (NAS, 1992, p. 13). The report went on to recommend that scientists and research institutions integrate into their curricula educational programs that foster integrity, and that institutions adopt formal guidelines on the conduct of responsible research.

A variety of other efforts have been made to foster integrity in the conduct of research. Professional groups, agencies of the federal government, and foundations have addressed the importance of integrity in the conduct of research for some time. For example, in 1982 the Association of American Medical Colleges published a policy statement entitled "The Maintenance of High Ethical Standards in the Conduct of Research," which emphasized the significance of integrity in the conduct of biomedical research.

Scientific societies, led by the American Association for the Advancement of Science (AAAS) and Sigma Xi, The Scientific Research Society, educate their members about the ethical issues associated with research and develop materials to help foster integrity in research (e.g., AAAS videos entitled "Integrity in Scientific Research"; see http://www.aaas.org/spp/video). AAAS has also joined forces with other scientific societies in considering ways to enhance the role of the societies in promoting integrity in research (AAAS, 2000). At the federal level, the National Science Foundation (NSF) and NIH have begun to support education in ethics and the responsible conduct of research (for examples see http://www.nsf.gov/sbe/ses/sdest; http://www.nih.gov/sigs/bioethics/researchethics.html).

In 2001, the Wellcome Trust, a private philanthropy in England, published draft guidelines for *Good Practice in Biomedical Research* (Wellcome Trust, 2001). Although these guidelines will apply only to those receiving funds from the Wellcome Trust, the guidelines have been looked upon as a positive development for research overall (Koenig, 2001).

Other organizations, such as Public Responsibility in Medicine and Research (http://www.primr.org), endeavor to promote the responsible conduct of research through broad educational efforts, such as national conferences and published reports.

Efficacies of Efforts to Foster Integrity

Opinions differ, and research is inconclusive, on which efforts to foster integrity are the most effective, but education consistently ranks high. As Ruth Fischbach noted during her testimony to this committee, "The elements of a research environment that promote integrity are education, education, education—ethical behavior favors a prepared mind." Both the 1989 and 1992 NAS reports stressed the importance of "educational programs that foster faculty and student awareness of concerns related to the integrity of the research process" (NAS, 1992, p. 13). ORI lists nine core content areas that it considers significant and that warrant inclusion in educational programs: (1) data acquisition, management, sharing, and ownership; (2) mentor-trainees relationships; (3) publication practices and responsible authorship; (4) peer review; (5) collaborative science; (6) human subjects; (7) research involving animals; (8) research misconduct; and (9) conflict of interest and commitment. These elements appear repeatedly in program and guidance documents (ORI, 2000).

Very little is known, however, about the efficacies of such educational programs as they are currently conducted. As noted earlier, the low incidence of allegations of misconduct means that these cases are not likely to be helpful in measuring changes in the research environment. Therefore, ORI is seeking guidance on what might be called "surrogate measures of the health of research environments." Examples of the kinds of measures that might be used include (1) the more familiar indicators of institutional effectiveness, such as the research record itself or the number of graduates who go on to prestigious positions; (2) outcome measures that an institution could use to assess the value added by educational efforts to promote abilities (e.g., moral reasoning) that relate to the responsible conduct of research; and (3) moral climate indicators, such as those designed to measure the organizational climate for business or the academic integrity of institutions of higher education. (See Appendix B for an extensive discussion of the outcome measures used to assess integrity in the research environment.)

To measure and monitor the climate of an institution, two kinds of indicators are used: the *perceptions of members* that the environment values and supports responsible conduct, discourages questionable practices, and censures misconduct (data can be collected by using surveys, interviews, or focus groups), and *process indicators*, such as the number and quality of improvements that an institution has made in the processes and procedures used to support research activities (including such things as the accreditation of an institution's institutional review board or the systems used to monitor the use of funds). Outcome measures that assess individual capacities could be used to assess the effectiveness of institutional efforts to promote the responsible conduct of individuals.

Current Research on Integrity in Research[2]

The 1992 COSEPUP report *Responsible Science: Ensuring the Integrity of the Research Process* (NAS, 1992) noted the lack of research-based knowledge about research misconduct. To a small but real extent, integrity in

---

[2]This section is based on a commissioned paper that David Guston prepared for the committee.

research has now become an object of research itself, and ORI has begun funding grants on "research on research integrity" (RRI). The first request for applications (RFA) in this area was issued in summer 2000. The National Institute of Neurological Disorders and Stroke (NINDS) and the National Institute of Nursing Research (NINR) joined ORI in the RRI grant program. In July 2001, ORI announced awards for seven 2-year grants, five funded by ORI and one funded by each of the collaborating institutes. Table 1-1 lists the topics of each of these grants. Although the topics are diverse, none of the currently funded projects specifically addresses ways to assess the effectiveness of interventions intended to foster integrity in research.

ORI, again in collaboration with NINDS and NINR, issued a second request for applications in May 2001. The RFA sought proposals that would "provide generalizable empirical knowledge about the ways in which researchers and research institutions meet or fail to meet their professional responsibilities in the conduct, evaluation, and reporting of research" (ORI, 2001b, p. 2). ORI also conducts research studies on its own. Two current studies of interest, to be released in 2003, are surveys of research integrity measures (not the integrity of research measures as stated) used in biomedical research laboratories and the incidence of research misconduct in biomedical research.

NSF, since 1989, has funded a small number of projects related to research misconduct. Tables 1-2 and 1-3 detail information about eight funded projects directly related to integrity in research and four funded projects indirectly related to integrity in research, respectively. During this period, NSF has spent just over $2 million on all of these projects together. Many of the funded grants relate to education and training rather than to empirical investigation into integrity or misconduct as phenomena in and of themselves.

TABLE 1-1 Grants Funded by ORI in the First Round of Research on Integrity in Research

Grant Title
Research Integrity in Pharmacological Clinical Trials
Perceived Organizational Justice in Scientific Dishonesty
Quality Assurance and Data in Clinical Trials
Data Sharing and Data Withholding Among Trainees in Science
Organizational Influences on Scientific Integrity
Work-Strain, Career Course, and Scientific Integrity
Management Decisions in Financial Conflicts of Interest

SOURCES: ORI (2001a) and M. Scheetz, Office of Research Integrity, Personal communication, October 10, 2001, with D. Guston, author of commissioned paper.

TABLE 1-2   NSF Awards Directly Related to Integrity in Research, 1989 to Present

Year	Amount ($)	Title
1991	36,000	Openness, Secrecy, Authorship, and Intellectual Property
1992	25,499	Planning Grant on Preserving Scientific Integrity in the Behavioral Sciences
1993	84,998	Sharing Research Data: An Examination of Practices
1994	22,500	Dynamic Issues in Scientific Integrity: Collaborative Research
1995	164,090	Professional Norms of Researchers in Cellular and Molecular Biology
1996	10,000	Conference: Historical Perspectives on Scientific Authorship
1998	171,261	Graduate Research Ethics Program
2001	763,907	Continuing and Expanding the Graduate Research Ethics Program

NOTE: Projects were identified on NSF's online awards database by searching on the terms "scientific misconduct" (yielding 5 hits) and "scientific integrity" (yielding 12 hits), although not all of the hits were fully relevant. Searching on "research integrity" and "research misconduct" yielded zero hits. Searching on the term "scientific ethics" yielded 42 hits, but there was no overlap with the "misconduct" and "integrity" hits, and so these were ignored.

ORI notes in the second RFA on research on research integrity that "no systematic effort has been made to evaluate different approaches to transmitting high standards for integrity in research, making it difficult to know which ones, if any, are effective" (ORI, 2001b, p. 1). The IOM committee's evaluation of the available literature supports this statement and further emphasizes the need for research in this area (see Chapter 7).

TABLE 1-3   NSF Awards Indirectly Related to Integrity in Research, 1989 to Present

Year	Amount ($)	Title
1989	600,255	Revision of the CHEM Study Films
1993	51,250	Undergraduate Research Participation: Collaborative Cross-Disciplinary Research in Biology
1996	11,858	Professional Development for Emerging Neuroscientists
1998	154,800	Research Experience for Undergraduates: Experimental Biology

NOTE: Projects were identified on NSF's online awards database by searching on the terms "scientific misconduct" (yielding 5 hits) and "scientific integrity" (yielding 12 hits), although not all of the hits were fully relevant. Searching on "research integrity" and "research misconduct" yielded zero hits. Searching on the term "scientific ethics" yielded 42 hits, but there was no overlap with the "misconduct" and "integrity" hits, and so these were ignored.

THE CHANGING RESEARCH ENVIRONMENT

The research environment changes continuously, and these changes influence the culture and conduct of research. For example, the once clear-cut lines between academic and for-profit research that existed during the growth in federal funding for medical and health-related research after World War II have become increasingly blurred. Industry expenditures for medical and health-related research conducted in the United States have been rising faster than federal-sector expenditures (NIH, 1996; NSF, 2000a). As a result, industry funding plays an important role in the conduct of medical and health-related research.

Because science is a cumulative, interconnected, and competitive enterprise, with tensions among the various societies in which research is conducted, now more than ever researchers must balance collaboration and collegiality with competition and secrecy. Another result of these tensions is conflict-of-interest and intellectual property issues, which are increasingly important to administrators of research institutions. Careful management of an institution's discoveries and developments can yield significant funding in the form of licensed patents, royalties, and investment by industries. Lack of careful control over intellectual property and potential conflicts of interest and commitment, however, can lead, at a minimum, to lost opportunities and, more seriously, to legal actions, loss of research funding, and other penalties. In addition, the institution's reputation may be tarnished and the institution may lose public and stakeholder trust. However, the management of conflicts of interest, although certainly important and desirable, does not ensure integrity in the conduct of research.

Because so much of modern life is based on advances in science and technology, these advances have generated large industries. Academic laboratories in the United States have become, in a sense, small businesses that are constantly seeking capital and the brightest minds to work in them (Grinnell, 1999b). In addition, information technologies contribute to all areas of research, while they simultaneously raise challenges in terms of data sharing, data protection, and personal privacy. As a result of the various elements and issues related to research and the research environment discussed above, running a laboratory requires more than the management of daily scientific activities. Success requires not only the intellectual ability to conduct research but also the capability to manage people and finances (and related conflicts of interest and commitment), adhere to regulations, and ensure such outputs as publications and intellectual property (Davis, 1999). As part of these responsibilities, research leaders and administrators should foster a climate that supports the responsible conduct of research. However, despite the importance of integrity to sound research, the means of promoting integrity in the individual researcher and developing an institutional climate that fosters integrity are not precisely known.

One of the more difficult situations that a research manager or administrator faces is how to handle situations of questionable research practices or outright misconduct. Unfortunately, fabrication, falsification, and plagiarism—even though they are relatively rare—garner significant attention. Although the federal government plays an important role in the management of misconduct in research, regulation alone cannot foster integrity. The commitment must come from individuals, the broad scientific community, and its institutions.

The "where," "what," and "who" in scientific research are very broad categories. Research is conducted everywhere, often collaboratively: in institutions of higher education, government facilities, and industry settings. Within each of these settings there can be multiple smaller units (e.g., departments, divisions within a department, research groups within a division). The research disciplines span the life, physical, earth, and social sciences. Each discipline has different cultures, populations, and "generations" of scientists, including students, trainees, junior faculty and researchers, tenured faculty, staff scientists, technical assistants, and administrators. The research enterprise includes not only those actively conducting research but also research sponsors, human research subjects, administrative and financial support staff, statisticians, animal handlers, intellectual property and business development managers, suppliers, manufacturers, professional organizations, publishers of scientific journals, and a host of other players. All of these players and their actions must be taken into account when considering the responsible conduct of research.

FOCUS OF THE REPORT

As with any system being scientifically examined, the research environment itself contains variables and constants. The most unpredictable and influential variable is the individual scientist. The human contribution to the research environment is greatly shaped by each individual's professional integrity, which in turn is influenced by the individual's educational background and cultural and ethical upbringing. These result in values and attitudes that contribute to the formation of the individual's identity, unique personality traits, and ethical decision-making abilities. Because each researcher brings unique qualities to the research environment, the constants must come from the environment itself. Research institutions should consistently and effectively provide training and education, policies and procedures, and tools and support systems. Institutional expectations should be unambiguous, and the consequences of each individual's conduct or misconduct should be clear. Anyone needing assistance should have ready access to knowledgeable leaders and be able to seek help without fear of retribution.

This report therefore focuses on the research environment and attempts to define and describe those elements that enable and encourage unique individuals, regardless of their role in the research organization or their backgrounds upon entry into that organization, to act with integrity. Although integrity and misconduct are related, the focus of this report is on integrity. The committee does not discuss or draw conclusions about current or proposed regulations or definitions relating to misconduct. The committee's goal was not to advocate any specific policy or process but, rather, was to define the desired outcome and to set forth a set of initiatives that it believes will increase the levels of integrity among individuals in research institutions. The committee considered approaches that could be used to foster integrity and methods that could be used to assess the effectiveness of those approaches. The majority of these approaches can and should be proactively initiated and administered by research institutions.

The committee's focus on the responsible conduct of research within academic institutions does not imply lack of interest in the environment for integrity in other research contexts, such as private research institutes or research at for-profit organizations. However, virtually all investigators begin their research careers in a university setting; therefore, the university research group can be considered the crucible for education in research. Additionally, as the principal recipients of public research funds, academic research groups have been the major focus of concern for integrity in research.

It should also be noted that this report emphasizes the education of trainees (graduate and medical students and postdoctoral fellows), not because the committee believes this is where a problem exists but, rather, because this is where the future lies. The committee hopes that focusing efforts on the next generation of researchers and scientific leaders will yield the greatest and most enduring change. This is not to say that senior researchers, faculty, and administrators cannot change or improve, and educational efforts should certainly be designed to reach all those involved in scientific research at all levels. As noted in a previous report, "the educational process should begin early in the training of future scientists and continue through the most senior stages" (DHHS, 1995, p. 15). The principles of adult learning and the discussion of the development of abilities that give rise to responsible conduct described in Chapter 5 are applicable to educational efforts at all levels of adult education and across all scientific disciples.

The committee acknowledges the difficulty of changing an organization's culture and ethical climate. The history of "reforms" in medical education has led to numerous reports, recommendations, and attempted reforms, which overall have amounted to "reform without change" (Bloom, 1988). One explanation is the focus on change in curriculum without concurrent change in the environment for teaching and learning. Bloom (1995, p. 907) notes that change "cannot be accomplished only by adding to or changing existing curriculum components. It must include a change in the teaching and learning environment, or in what we mean by socialization for a profession." Teaching of values is important, but "only when students can see those values operative within their schools' service and research programs will the lesson become fully effective" (Bloom, 1995, p. 908). As such, the committee discusses education in the responsible conduct of research as an integral component of conducting scientific research (Chapter 5).

For its evaluation of the elements that are most likely to foster integrity, the committee analyzed the available literature; consulted with experts in science, education, and organizational development; and commissioned papers (see Appendix A). In looking for evidence to inform its deliberations, the committee focused on what is known about the assessment of a moral climate and what is known about the effectiveness of attempts to teach responsible conduct. The committee members also drew heavily from their own experiences, those of their institutions, and those of invited experts.

ORGANIZATION OF THE REPORT

Following this introductory chapter, a discussion of assessing integrity in the research environment begins in Chapter 2 with the committee's definition of integrity as it applies to both individuals and institutions. The committee describes practices that embody these definitions. Chapter 3 presents an organizational framework for the research environment and discusses some of the key elements in more detail, including education, policies, culture, and assessment and quality improvement. In Chapter 4, the committee presents elements that in combination can foster integrity in the research environment and discusses the advantages and disadvantages of each. Chapter 5 expands the discussion of education as an important element in fostering integrity. Chapter 6 furthers the discussion of self-assessment as a preferred method for assessing integrity in the research environment. Concluding remarks, recommendations, and areas for further research are presented in Chapter 7.

The report includes five supplementary appendixes to provide the reader with additional information: Appendix A discusses the committee's data sources and summarizes the findings of its literature review. Appendix B describes existing outcome measures that might be used as a framework to develop instruments to assess integrity in the research environment. Appendix C provides a brief historical overview of integrity and misconduct over the past 10 years. Appendix D lists additional reading and resources. Biographical information on the committee members is presented in Appendix E.

Table 1-4 lists each of the individual tasks that the committee ad-

TABLE 1-4   Addressing the Charge

Task	Committee Action
Define the concept "research integrity."	**Chapter 2** describes integrity in research as it relates to both individual researchers and the institutions in which they work.
Describe and define the concept "research environment."	**Chapter 3** uses an open-systems model (often used to describe social organizations) to provide a general framework within which the "research environment" can be understood.
Identify elements of the research environment that promote integrity in research.	**Chapters 2 and 3** identify the elements within a research organization that are relevant to integrity in research. **Chapter 5** expands upon the discussion of education as an important element.
Indicate how the elements may be measured. *Suggest an appropriate methodology for collection of the data.* *Cite appropriate outcome measures.*	**Chapter 6** describes an approach for evaluation of the environment for integrity in research based on methods of self-assessment and peer review that are incorporated into existing processes for accreditation of educational and research institutions. **Appendix B** presents examples of the types of instruments that could be used to collect data as part of a self-assessment and examples of elements and outcomes assessed in other models.
Make recommendations regarding the adoption and implementation by research institutions, government agencies, scientific societies, and others (as appropriate) of those elements of the research environment identified to promote integrity in research.	**Chapter 4** presents three approaches that represent alternative ways of influencing behavior and improving outcomes and the strengths and weaknesses of each. **Chapters 5 and 6** discuss the implementation of education in responsible conduct and evaluation by self-assessment, respectively. **Chapter 7** includes recommendations and identifies areas in which more research is needed to further identify, characterize, and measure elements of the research environment that promote integrity.
Convene a public meeting to discuss the IOM report, its recommendations, and potential strategies for their implementation.	A public meeting will be held in the fall of 2002.

BOX 1-1
Glossary of Terms Used in This Report

educational programs: programs that aim to develop students mentally, morally, or aesthetically through instruction

ethical (or moral) climate: the prevailing moral beliefs that provide the context for conduct (i.e., the prescribed behaviors, beliefs, and attitudes within the community and the sanctions expressed); the stable, psychologically meaningful, and shared *perceptions* of organizational members are used as indicators of climate

ethical (or moral) reasoning: the ability to examine systematically the ethical dimensions of a situation and then choose and defend a position on the issue on ethical or moral grounds

ethics: principles of character often believed to transcend particular communities

morals*: normative principles of right or wrong in behavior adopted within particular communities

organizational culture: the set of shared norms, values, beliefs, and assumptions along with the behavior and other artifacts that express these orientations—including symbols, rituals, stories, and language

research: systematic investigation or experimentation aimed at generating generalizable information and knowledge

research environment: the combined social and cultural conditions that influence the life of an individual investigator, research unit, or research institution

research institution: all organizations conducting research, including, for example, colleges and universities, intramural federal research laboratories, federally funded research and development centers, industrial laboratories, and other research institutes

science: knowledge or a system of knowledge covering general "truths" or the operation of general laws, especially those obtained and tested through the scientific method

training programs: programs that provide a set of skills and experiences

---

*Although *morals* and *ethics* have different meanings as technical terms, they are often used interchangeably.

dresses in this report and the chapter(s) that contains the majority of the committee's response to them. The committee notes that tasks four, five, and six are particularly demanding. Professional consulting firms can spend years, and a significant budget, developing and validating assessment instruments. As such, the committee was not equipped to recommend specific methods and measures. Instead, the committee recommends an overall approach—institutional self-assessment followed by peer review, preferably as part of the standard accreditation process—as a means to collect and assess data (Chapter 6). The committee also provides information on relevant assessment tools that might be adapted to the research environment (Appendix B). As noted throughout the report, empirical data on evaluating ethical climate before and after implementation of specific practices or policies are lacking, and the committee believes that it went as far as the current data would responsibly allow in making its recommendations.

REFERENCES

AAAS (American Association for the Advancement of Science). 2000. *The Role and Activities of Scientific Societies in Promoting Research Integrity.* A report of a conference, April 10, 2000, Washington, DC. [Online]. Available: http://www.aaas.org/spp/dspp/sfrl/projects/integrity.htm [Accessed January 7, 2002].

American Association of Medical Colleges. 1982. The maintenance of high ethical standards in the conduct of research. A policy statement. American Association of Medical Colleges, Washington, DC.

Bloom SW. 1988. Structure and ideology in medical education: An analysis of resistance to change. *Journal of Health and Social Behavior* 29:294–306.

Bloom SW. 1995. Editorial: Reform without change? Look beyond the curriculum. *American Journal of Public Health* 85:907–908.

Davis B. 1991. Is the Office of Scientific Integrity too zealous? *The Scientist* 5 (10):12.

Davis TP, ed. 1999. *Management of Biomedical Research Laboratories.* Proceedings of a national conference. October 1–3, 1999, The University of Arizona, Tucson.

DHHS (U.S. Department of Health and Human Services). 1995. *Integrity and Misconduct in Research: Report of the Commission on Research Integrity.* Rockville, MD: Office of the Secretary, Office of Research Integrity, DHHS.

DHHS. 2000. Statement of Organization, Functions, and Delegations of Authority. *Federal Register* 65:30600–30601.

Grinnell F. 1992. *The Scientific Attitude,* 2nd ed. New York, NY: The Guilford Press.

Grinnell F. 1999a. Ambiguity, trust, and responsible conduct of research. *Science and Engineering Ethics* 5:205–214.

Grinnell F. 1999b. Responsibility, conflict and ambiguity in the management of biomedical research laboratories. In: Davis TP, ed. *Management of Biomedical Research Laboratories.* Tucson, AZ: University of Arizona. Pp. 29–35.

Guston DH. 2000. *Between Politics and Science: Assuring the Integrity and Productivity of Research.* New York, NY: Cambridge University Press.

Hamilton DP. 1991. Can OSI withstand a scientific backlash? *Science* 253:1084–1086.

IOM (Institute of Medicine). 2001. *Preserving Public Trust.* Washington, DC: National Academy Press.

Koenig R. 2001. Wellcome rules widen the net. *Science* 293:1411–1412.

Macrina FL. 2000. *Scientific Integrity: An Introductory Text with Cases,* 2nd ed. Washington, DC: ASM Press.

NAS (National Academy of Sciences). 1989a. *On Being a Scientist.* Washington, DC: National Academy Press.

NAS. 1989b. *The Responsible Conduct of Research in the Health Sciences.* Washington, DC: National Academy Press.

NAS. 1992. *Responsible Science: Ensuring the Integrity of the Research Process,* Vol. 1. Washington, DC: National Academy Press.

NAS. 1993. *Responsible Science: Ensuring the Integrity of the Research Process,* Vol. 2. Washington, DC: National Academy Press.

NAS. 1995. *On Being a Scientist,* 2nd ed. Washington, DC: National Academy Press.

NIH (National Institutes of Health). 1996. *Estimates of National Support for Health R&D by Source or Performer, FY 1986–1995.* [Online]. Available: http://grants.nih.gov/grants/award/trends96/pdfdocs/FEDTABLA.PDF [Accessed December 6, 2001].

NSF (National Science Foundation). 2000a. U.S. and international research and development: funds and alliances. In: *Science and Engineering Indicators-2000.* Arlington, VA: NSF. [Online]. Available: http://www.nsf.gov/sbe/srs/seind00/start.htm [Accessed December 5, 2001].

NSF. 2000b. Science and Technology: Public Attitudes and Public Understanding. In: *Science and Engineering Indicators-2000.* Arlington, VA: NSF. [Online]. Available: http://www.nsf.gov/sbe/srs/seind00/start.htm [Accessed September 28, 2001].

ORI (Office of Research Integrity). 1999. Review group recommendations being implemented rapidly. *ORI Newsletter* 8(1):5–6.

ORI. 2000. *PHS Policy on Instruction in the Responsible Conduct of Research (RCR).* [Online]. Available: http://ori.dhhs.gov/html/programs/finalpolicy.asp [Accessed November 2001].

ORI. 2001a. Seven Studies Funded by Research Integrity Program. *ORI Newsletter* 9(4):1, 4.

ORI. 2001b. Research on Research Integrity. *NIH Guide.* RFA-NS-02-005. [Online] Available: http://grants2.nih.gov/grants/guide/rfa-files/RFA-NS-02-005.html [Accessed March 18, 2002].

OSTP (Office of Science and Technology Policy). 2000. Federal policy on research misconduct; Preamble for research misconduct policy. Notification of Final Policy. *Federal Register* 65:76260–76264.

Wellcome Trust. 2001. *Good Practice in Biomedical Research.* Draft guidelines. [Online]. Available: http://www.wellcome.ac.uk/en/1/awtvispolgrp.html [Accessed March 14, 2002].

Yarborough M, Sharp RR. 2002. Restoring and preserving trust in biomedical research. *Academic Medicine* 77:8–14.

2

Integrity in Research

The pursuit and dissemination of knowledge enjoy a place of distinction in American culture, and the public expects to reap considerable benefit from the creative and innovative contributions of scientists. As science becomes increasingly intertwined with major social, philosophical, economic, and political issues, scientists become more accountable to the larger society of which they are a part. As a consequence, it is more important than ever that individual scientists and their institutions periodically reassess the values and professional practices that guide their research as well as their efforts to perform their work with integrity.

Society's confidence in and support of research rest in large part on public trust in the integrities of individual researchers and their supporting institutions. The National Academies' report *On Being a Scientist* states: "The level of trust that has characterized science and its relationship with society has contributed to a period of unparalleled scientific productivity. But this trust will endure only if the scientific community devotes itself to exemplifying and transmitting the values associated with ethical scientific conduct" (NAS, 1995, preface). It is therefore incumbent on all scientists and scientific institutions to create and nurture a research environment that promotes high ethical standards, contributes to ongoing professional development, and preserves public confidence in the scientific enterprise (Grinnell, 1999; IOM, 2001; Resnik, 1998; Yarborough and Sharp, 2002).

Government oversight of scientific research is important, but such oversight, often in the form of administrative rules, typically stipulates what cannot be done; it rarely prescribes what should be done (see Chapter 4 for further discussion of the strengths and limitations of a regulatory approach). In essence, government rules define the floor of expected behavior. More, however, should be expected from scientists when it comes to the responsible conduct of research. By appealing to the conscience of individual scientists, the scientific community as a whole should seek to evoke the highest possible standard of research behavior. When institutions committed to promoting integrity in research support those standards, the likelihood of creating an environment that advances responsible research practices is greatly enhanced. It is essential that institutions foster a culture of integrity in which students and trainees, as well as senior researchers and administrators, have an understanding of and commitment to integrity in research.

The committee's task was to define integrity for the particular activity of research as conducted within contemporary society. Integrity has two general senses. The first sense concerns wholeness; the second, soundness of moral principle (*Oxford English Dictionary*, 1989). Plato and subsequent philosophers have argued that leading the good life depends on a person's success in integrating moral, religious, and philosophical convictions. In conversations with experts in ethics and others, the committee found no consensus regarding whether a person could exhibit high integrity in research but not in other aspects of his life. Consequently, the committee decided to focus on the second aspect of integrity—namely, soundness of moral principle in the specific context of research practice.

INTEGRITY IN RESEARCH

Integrity characterizes both individual researchers and the institutions in which they work. For individuals, it is an aspect of moral character and experience.[1] For institutions, it is a matter of creating an environment that promotes responsible conduct by embracing standards of excellence, trustworthiness, and lawfulness that inform institutional practices.

For the individual scientist, integrity embodies above all a commitment to intellectual honesty and personal responsibility for one's actions and to a range of practices that characterize responsible research conduct. These practices include:

- intellectual honesty in proposing, performing, and reporting research;

---

[1] Further discussion of moral character and behavior and the development of abilities that give rise to responsible conduct can be found in Chapter 5.

- accuracy in representing contributions to research proposals and reports;
- fairness in peer review;
- collegiality in scientific interactions, including communications and sharing of resources;
- transparency in conflicts of interest or potential conflicts of interest;
- protection of human subjects in the conduct of research;
- humane care of animals in the conduct of research; and
- adherence to the mutual responsibilities between investigators and their research teams.

Individual scientists work within complex organizational structures. (These structures and their interactions are described in detail in Chapter 3.) Factors that promote responsible conduct can exert their influences at the level of the individual; at the level of the work group (e.g., the research group); and at the level of the research institution itself. These different organizational levels are interdependent in the conduct of research. Institutions seeking to create an environment that promotes responsible conduct by individual scientists and that fosters integrity must establish and continuously monitor structures, processes, policies, and procedures that:

- provide leadership in support of responsible conduct of research;
- encourage respect for everyone involved in the research enterprise;
- promote productive interactions between trainees and mentors;
- advocate adherence to the rules regarding all aspects of the conduct of research, especially research involving human subjects and animals;
- anticipate, reveal, and manage individual and institutional conflicts of interest;
- arrange timely and thorough inquiries and investigations of allegations of scientific misconduct and apply appropriate administrative sanctions;
- offer educational opportunities pertaining to integrity in the conduct of research; and
- monitor and evaluate the institutional environment supporting integrity in the conduct of research and use this knowledge for continuous quality improvement.

Leadership by individuals of high personal integrity helps to foster an environment in which scientists can openly discuss responsible research practices in the face of conflicting pressures. All those involved in the research enterprise should acknowledge that integrity is a key dimension of the essence of being a scientist and not a set of externally imposed regulatory constraints.

INTEGRITY OF THE INDIVIDUAL SCIENTIST

As noted above, the committee has identified a range of key practices that pertain to the responsible conduct of research by individual scientists. The following sections elucidate the practices.[2]

Intellectual Honesty in Proposing, Performing, and Reporting Research

Intellectual honesty in proposing, performing, and reporting research refers to honesty with respect to the meaning of one's research. It is expected that researchers present proposals and data honestly and communicate their best understanding of the work in writing and verbally. The descriptions of an individual's work found in such communications frequently present selected data from the work organized into frameworks that emphasize conceptual understanding rather than the chronology of the discovery process. Clear and accurate research records must underlie these descriptions, however. Researchers must be advocates for their research conclusions in the face of collegial skepticism and must acknowledge errors.

Accuracy in Representing Contributions to Research Proposals and Reports

Accuracy in representing one's contributions to research proposals and reports requires the assignment of credit. It is expected that researchers will not report the work of others as if it were their own. This is plagiarism. Furthermore, they should be honest with respect to the contributions of colleagues and collaborators. Decisions regarding authorship are best anticipated at the outset of projects rather than at their completion. In publications, it should be possible in principle to specify each author's contribution to the work. It also is expected that researchers honestly acknowledge the precedents on which their research is based.

---

[2]See the section of Appendix D entitled Responsible Scientific Conduct for resources with case studies that can be used in a teaching setting to further illustrate the topics discussed here.

Fairness in Peer Review

Fairness in peer review means that researchers should agree to be peer reviewers only when they can be impartial in their judgments and only when have revealed their conflicts of interest. Peer review functions to maintain the excellence of published scientific work and ensure a merit-based system of support for research. A delicate balance pervades the peer-review system, because the best reviewers are precisely those individuals who have the most to gain from "insider information": they are doing similar work and they will be unable to "strike" from memory and thought what they learn through the review process. Investigators serving as peer reviewers should treat submitted manuscripts and grant applications fairly and confidentially and avoid using them inappropriately.

Collegiality in Scientific Interactions, Including Communications and Sharing of Resources

Collegiality in scientific interactions, including communications and sharing of resources requires that investigators report research findings to the scientific community in a full, open, and timely fashion. At the same time, it should be recognized that the scientific community is highly competitive. The investigator who first reports new and important findings gets credited with the discovery.

It is not obvious that rapid reporting is the approach that is always the most conducive to progress. Intellectual property provisions and secrecy allow for patents and licensure and encourage private investment in research. Furthermore, even for publicly funded research, a degree of discretion may permit a research group to move ahead more efficiently. Conversely, an investigator who delays reporting important new findings risks having others publish similar results first and receiving little recognition for the discovery. Knowing when and how much to tell will always remain a challenge in scientific communication.

Once scientific work is published, researchers are expected to share unique materials with other scientists in a reasonable fashion to facilitate confirmation of their results. (The committee recognizes that there are limits to sharing, especially when doing so requires a time or cost commitment that interferes with the function of the research group.) When materials are developed through public funding, the requirement for sharing is even greater. Public funding is based on the principle that the public good is advanced by science conducted in the interest of humanity. This commitment to the public good implies a responsibility to share materials with others to demonstrate reproducibility and to facilitate the replication and validation of one's work by responding constructively to inquiries from other scientists, particularly regarding methodologies.

Collegiality and sharing of resources is also an important aspect of the interaction between trainees and their graduate or postdoctoral advisers. Students and fellows will ultimately depart the research team, and discussion of and planning for departure should occur over the course of their education. Expectations about such issues as who inherits intellectual property rights to a project or to the project itself upon the trainee's departure should be discussed when the trainee first joins the research group and should be revisited periodically over the course of the project (NAS, 2000).

Transparency in Conflicts of Interest or Potential Conflicts of Interest

A conflict of interest in research exists when the individual has interests in the outcome of the research that may lead to a personal advantage and that might therefore, in actuality or appearance, compromise the integrity of the research. The most compelling example is competition between financial reward and the integrity of the research process. Religious, political, or social beliefs can also be undisclosed sources of research bias.

Many scientific advances that reach the public often involve extensive collaboration between academia and industry (Blumenthal et al., 1996; Campbell et al., 1998; Cho et al., 2000). Such collaborations involve consulting and advisory services as well as the development of specific inventions, and they can result in direct financial benefit to both individuals and institutions. Conflicts of interest reside in a situation itself, not in any behavior of members of a research team. Thus, researchers should disclose all conflicts of interest to their institutions so that the researchers and their work can be properly managed. They should also voluntarily disclose conflicts of interest in all publications and presentations resulting from the research. The committee believes that scientific institutions, including universities, research institutes, professional societies, and professional and lay journals, should embrace disclosure of conflicts of interest as an essential component of integrity in research.

Protection of Human Subjects in the Conduct of Research

The protection of individuals who volunteer to participate in research is essential to integrity in research. The ethical principles underlying such research have been elaborated on in international codes and have been integrated into national regulatory frameworks (in the United States, 45 C.F.R. § 46, 2001). Elements included in such frameworks pertain to the quality and importance of the science, its risks and benefits, fairness in the selection of subjects, and, above all, the voluntary participation and informed consent of subjects. To ensure the conformance of research efforts with these goals, institutions carry out extensive research subject protection programs. To be successful, such programs require high-level, functioning institutional review boards, knowledgeable investigators, ongoing performance assessment through monitoring and feedback, and educational programs (IOM, 2001). The IOM report *Preserving Public Trust* (IOM, 2001) focuses specifically on the important topic of research involving human subjects, and further discussion is not included here.

Humane Care of Animals in the Conduct of Research

The humane care of animals is essential for producing sound science and its social benefits. Researchers have a responsibility to engage in the humane care of animals in the conduct of research. This means evaluating the need for animals in any particular protocol, ensuring that research animals' basic needs for life are met prior to research, and carefully considering the benefits of the research to society or to animals versus the likely harms to any animals included as part of the research protocol. Procedures that minimize animal pain, suffering, and distress should be implemented. Research protocols involving animals must be reviewed and approved by properly constituted bodies, as required by law (Animal Welfare Act of 1966 [PL 89-544], inclusive of amendments passed in 1970 [PL 91-579], 1976 [PL 94-279], 1985 [PL 99-198], and 1990 [PL 101-624] and subsequent amendments) and professional standards (AAALAC, 2001; NRC, 1996).

Adherence to the Mutual Responsibilities Between Investigators and Their Research Teams

Adherence to the mutual responsibilities between investigators and members of their research teams refers to both scientific and interpersonal interactions. The research team might include other faculty members, colleagues (including coinvestigators), and trainees (undergraduate students, graduate and medical students, postdoctoral fellows), as well as employed staff (e.g., technicians, statisticians, study coordinators, nurses, animal handlers, and administrative personnel). The head of the research team should encourage all members of the team to achieve their career goals. The interpersonal interactions should reflect mutual respect among members of the team, fairness in assignment of responsibilities and effort, open and frequent communication, and accountability. In this regard, scientists should also conduct disputes professionally (Gunsalus, 1998). (The American Association of University Professors (AAUP) guidelines on academic freedom and professional ethics articulate the obligation of members of the academic community to root their statements in fact and to respect the opinions of others [AAUP, 1987, 1999].)

Mentoring and Advising

    Mentor is often used interchangeably with *faculty adviser*. However, a mentor is more than a supervisor or an adviser (Bird, 2001; Swazey and Anderson, 1998).[3] An investigator or research adviser may or may not be a mentor. Some advisers may be accomplished researchers but do not have the time, training, or ability to be good mentors (NAS, 2000). For a trainee, "a mentoring relationship is a close, individualized relationship that develops over time between a graduate student (or other trainee) and a faculty member (or others) that includes both caring and guidance" (University of Michigan, 1999, p. 5). A successful mentoring relationship is based on mutual respect, trust, understanding, and empathy (NAS, 1997). Mentoring relationships can extend throughout all phases of a science career, and, as such, they are sometimes referred to as mentor-protégé or mentor-apprentice relationships, rather than mentor-trainee relationships.

    The committee believes that *mentor* should be the dominant and usual role of the laboratory director or research advisor in regard to his or her trainee. With regard to such mentor-trainee relationships, responsibilities include a commitment to continuous education and guidance of trainees, appropriate delegation of responsibility, regular review and constructive appraisal of trainees, fair attribution of accomplishment and authorship, and career guidance, as well as help in creating opportunities for employment and funding. For the trainee, essential elements include respect for the mentor, loyalty to the research group, a strong commitment to science, dedication to the project, careful performance of experiments, precise and complete record keeping, accurate reporting of results, and a commitment to oral and written presentations and publication. It should be noted that most academic research institutions play a dual role. On the one hand, they are concerned with producing original research; on the other, with educating students. The two goals are compatible, but when they come in conflict, it is important that the educational needs of the students not be forgotten. If students are exploited, then they will learn by example that such behavior is acceptable.

---

[3]A special issue of *Science and Engineering Ethics* (7:451–640, 2001) is devoted to the relationship between mentoring and responsible conduct.

SUPPORT OF INTEGRITY BY THE RESEARCH INSTITUTION

The individual investigator and the laboratory or research unit carry out their functions in institutions that are responsible for the management and support of the research carried out within their domains. The institutions, in turn, are regulated by governmental and other bodies that impose rules and responsibilities (see Chapter 3 for further discussion). The vigor, resources, and attitudes with which institutions carry out their responsibilities will influence investigators' commitment and adherence to responsible research practices.

Provide Leadership in Support of Responsible Conduct of Research

It takes the leadership of an institution to promulgate a culture of responsible research. This involves the development of a vision for the research enterprise and a strategic plan. It is the responsibility of the institution leadership to develop programs to orient new researchers to institutional policies, rules, and guidelines; to sponsor opportunities for dialogue about new and emerging issues; and to sponsor continuing education about new policies and regulations as they are developed. Furthermore, institutional leaders have the responsibility to ensure that such programs are carried out, with appropriate delegation of responsibility and accountability and with adequate resources.

The observed actions of institutions in problem situations communicate as strongly (or perhaps more strongly) about responsible conduct as do any policies or programs. Institutional leaders (e.g., chancellor, president, dean, CEO) set the tone for the institutions with their own actions. Research leaders should set an example not only in their own research practices but also in their willingness to engage in dialogue about ethical questions that arise (Sigma Xi, 1999). McCabe and Pavela note that "faculty members who seek to instill a sense of social obligation without affirming personal virtues are planting trees without roots" (McCabe and Pavela, 1998, p.101).

Encourage Respect for Everyone Involved in the Research Enterprise

An environment that fosters competence and honest interactions among all participants in the investigative process supports the integrity of research. Institutions have many legally mandated policies that foster mutual respect and trust—for example, policies concerning harassment, occupational health and safety, fair employment practices, pay and benefits, protection of research subjects, exposure to ionizing radiation, and due process regarding allegations of research misconduct. State and local policies and guidelines governing research may be in effect as well. It is anticipated that through a process of self-assessment, institutions can identify issues and develop programs that further integrity in research (see Chapter 6 for further discussion). Fair enforcement of all institutional policies is a critical element of the institutional commitment to integrity in research. That is not enough, however.

Support Systems

Within the research institution, there can be multiple smaller units (e.g., departments, divisions within a department, research groups within a division). Within these institutional subunits, there will always be power differences between members of the group. Consequently, research institutions require support mechanisms—for example, ombudspersons—that research team members can turn to for help when they feel they are being treated unfairly. Institutions need to provide guidance and recourse to anyone with concerns about research integrity (e.g., a student who observes a lack of responsible conduct by a senior faculty member). Support systems should be accessible (multiple entry points for those with questions) and have a record of reaching objective, fact-based decisions untainted by personal bias or conflicts of interest (Gunsalus, 1993). Lack of recourse within the institution for those individuals who have concerns about possible misconduct will undermine efforts to foster a climate of integrity. Equally important to having support systems in place is the dissemination of information on how and where individuals may seek such support.

The ultimate goal for institutions should be to create a culture within which all persons on a research team can work effectively and realize their full potential.

Promote Productive Interactions Between Trainees and Mentors

Mentors play a special role in the development of new scientists. A mentor must consider the student's core interests and needs in preference to his or her own. Trainees and mentors are codependent and, at times, competitive. Trainees depend on their mentors for scientific education and training, for support, and, eventually, for career guidance and references. Mentors tend to be role models as well. Mentors depend on trainees for performing work and bringing fresh ideas and approaches to the research group. They can enhance the mentor's reputation as a teacher and as an investigator. Institutions should establish programs that foster productive relations between mentors and trainees, including training in mentoring and advising for faculty. Moreover, institutions should work to ensure that trainees are properly paid, receive reasonable benefits (including health insurance), and are protected from exploitation. Written guidelines, ombudspersons, and mutual evaluations can help to reduce problems and identify situations requiring remediation. As mentioned earlier in this chapter, the dual role academic research institutions play in both producing original research and educating students can be balanced, but when they come in conflict, educational interests of the student should take precedence.

Advocate Adherence to the Rules Regarding All Aspects of the Conduct of Research, Especially Research Involving Human Subjects and Animals

Effective advocacy by an institution of the rules involving the use of human subjects and animals in research involves much more than simply posting the relevant federal, state, and local regulations and providing "damage control" and formal sanctions when irregularities are discovered. At all levels of the institution, including the level of the dean, department chair, research group leader, and individual research group member, regular affirmation of the guiding principles underlying the rules is essential. The goal is to create an institutional climate such that anyone who violates these guiding principles through words or deeds is immediately made aware of the behavior and, when indicated, appropriately sanctioned.

Anticipate, Reveal, and Manage Individual and Institutional Conflicts of Interest

Research institutions must conduct their work in a manner that earns public trust. To do so, they must be sensitive to any conflict of interest that might affect or appear to affect their decisions and behavior in ways that could compromise their roles as trustworthy sources of information and policy advice or their obligations to ensure the protection of human research subjects. As research partnerships between industry and academic institutions continue to expand, with the promise of considerable public benefit, the management of real or perceived conflicts of interest in research requires that institutions have a written policy on such conflicts. The policy should apply to both institutions and individual investigators.

Institutional Conflicts of Interest

Institutions should have clearly stated policies and procedures by which they will guard against compromise by external influences. As with individual conflicts of interest, institutional leadership is not in the best position to determine whether a particular arrangement represents an unacceptable or manageable conflict of interest. Institutions should draw on independent reviews by external bodies and should have appropriate procedures for such reviews. Factors of concern include not only direct influences on institutional policy but also indirect influences on the use of resources, educational balance, and hiring of faculty, for example (AAU, 2001).

Institutional Responsibility for Investigator Conflicts of Interest

The policy on conflicts of interest should apply to individuals who are directly involved in the conduct, design, and review of research, including faculty, trainees, students, and administrators, and should clearly state their disclosure responsibilities. The policy should define conflicts of interest and should have means to convey an understanding of the term to the parties involved. It should delineate the activities and the levels and kinds of research-related financial interests that are and are not permissible, as well as those that require review and approval. The special circumstances associated with research involving human subjects should be specifically addressed. Beyond meeting their responsibility to ensure the dissemination and understanding of their policies, institutions should develop means to monitor compliance equitably. Detailed descriptions of institutional responsibilities in this area were recently reported by the Association of American Universities (AAU, 2001) and the Association of American Medical Colleges (AAMC, 2001), as described in Box 2-1.

BOX 2-1
Definition of Institutional Conflict of Interest

An "institutional conflict of interest . . . may occur when the institution, any of its senior management or trustees, or a department, school, or other sub-unit, or an affiliated foundation or organization, has an external relationship or financial interest in a company that itself has a financial or other interest in a faculty research project. Senior managers or trustees may also have conflicts when they serve on the boards of (or otherwise have an official relationship with) organizations that have significant business relationships with the university. The existence (or appearance) of such conflicts can lead to actual bias, or suspicion about possible bias, in the review or conduct of research at the university. If they are not evaluated or managed, they may result in choices or actions that are incongruent with the missions, obligations, or the values of the university" (AAU, 2001, p. i).

Arrange Timely and Thorough Inquiries and Investigations of Allegations of Scientific Misconduct and Apply Appropriate Sanctions

Every institution receiving federal funds for research and related activities must have in place policies and procedures for responding to allegations of research misconduct (42 C.F.R. § 50, §§ A, 1989; 45 C.F.R. § 689, 1996). Although the federal government imposes these requirements, the institutions must implement them. Their effectiveness depends on investigation of allegations of misconduct with vigor and fairness. The institution should embrace the notion that it is important to the quality and integrity of science that individuals report possible research misconduct. Means of protecting any individual who reports possible misconduct in good faith must be instituted.

In carrying out their responsibilities, institutions must ensure that faculty, students, and staff are properly informed of their rights and responsibilities. Those likely to receive allegations—for example, administrators, department chairs, and research group chiefs—must be fully informed of institutional provisions and trained in dealing with issues related to research conduct or misconduct. Mechanisms must be in place to protect the public's interest in the research record, the research subjects' health, and the financial interests of the institution, as well as to ensure notification of appropriate authorities. Clear lines of authority for management of the institution's response must exist, and, where indicated, appropriate sanctions should be applied or efforts should be made to protect or restore the reputations of innocent parties.

Offer Educational Opportunities Pertaining to Integrity in the Conduct of Research

Research institutions should provide students, faculty, and staff with educational opportunities related to the responsible conduct of research. These are mandatory for those involved in clinical research (NIH, 2000) and for recipients of Public Health Service training grants (NIH, 1989). These offerings should encourage open discussion of the values at stake and the ethical standards that promote responsible research practices. The core objective of such education is to increase participants' knowledge and sensitivity to the issues associated with integrity in research and to improve their ability to make ethical choices. It should give them an appreciation for the diversity of views that may be brought to bear on issues, inform them about the institutional rules and government regulations that apply to research, and instill in them the scientific community's expectations regarding proper research practice. Educational offerings should be flexible in their approach and be cognizant of normative differences among disciplines. Such programs should offer opportunities for the participants to explore the underlying values that shape the research enterprise and to analyze how those values are manifested in behaviors in different research environments

It is expected that effective educational programs will empower individual researchers, students, and staff in making responsible choices in the course of their research. Regular evaluation and improvement of the educational and behavioral effectiveness of these educational offerings should be a part of an institutional assessment. (See Chapter 5 for further discussion of education in the responsible conduct of research.)

Monitor and Evaluate the Institutional Environment Supporting Integrity in the Conduct of Research and Use This Knowledge for Continuous Quality Improvement

The main thrust of this report reflects the need for continuing attention toward sustaining and improving a culture of integrity in research. This requires diligent oversight by institutional management to ensure that the practices associated with integrity described above are carried out. It also requires examination of the policy-making process, the policies themselves, their execution, and the degree to which they are understood and adhered to by those affected. If researchers and administrators believe that the rules are excellent and that the institution applies them equitably, then the institutional commitment to integrity will be clear. Chapter 6 addresses ways to help identify those elements critical to establishment of the perception of moral commitment and determination of whether such commitments have been made.

SUMMARY

The committee believes that integrity in research is essential for maintaining scientific excellence and keeping the public's trust. The concept of integrity in research cannot, however, be reduced to a one-line definition. For a scientist, integrity embodies above all the individual's commitment to intellectual honesty and personal responsibility. It is an aspect of moral character and experience. For an institution, it is a commitment to creating an environment that promotes responsible conduct by embracing standards of excellence, trustworthiness, and lawfulness and then assessing whether researchers and administrators perceive that an environment with high levels of integrity has been created. This chapter has described multiple practices that are most likely to promote responsible conduct. Individuals and institutions should use these practices with the goal of fostering a culture in which high ethical standards are the norm, ongoing professional development is encouraged, and public confidence in the scientific enterprise is preserved.

REFERENCES

AAALAC (Association for Assessment and Accreditation of Laboratory Animal Care). 2001. *AAALAC International Rules of Accreditation.* [Online]. Available: http://www.aaalac. org/rules.htm [Accessed January 31, 2002].

AAMC (Association of American Medical Colleges). 2001. *Protecting Subjects, Preserving Trust, Promoting Progress.* [Online] Available: http://www.aamc.org/coitf [Accessed December 18, 2001].

AAU (Association of American Universities). 2001. *Report on Individual and Institutional Conflict of Interest.* [Online] Available: http://www.aau.edu/research/conflict.html [Accessed January 31, 2002].

AAUP (American Association of University Professors). 1987. *Statement on Professional Ethics.* [Online]. Available: http://www.aaup.org/statements/Redbook/Rbethics.htm [Accessed May 14, 2002].

AAUP. 1999. *Recommended Institutional Regulations on Academic Freedom and Tenure.* [Online]. Available: http://www aaup.org/statements/Redbook/Rbrir.htm [Accessed May 14, 2002].

Bird SJ. 2001. Mentors, advisors and supervisors: Their role in teaching responsible research conduct. *Science and Engineering Ethics* 7:455–468.

Blumenthal D, Causino N, Campbell E, Seashore Louis K. 1996. Relationships between academic institutions and industry in the life sciences: An industry survey. *New England Journal of Medicine* 334:368–373.

Campbell EG, Seashore Louis K, Blumenthal D. 1998. Looking a gift horse in the mouth. Corporate gifts supporting life sciences research. *Journal of the American Medical Association* 279:995–999.

Cho MK, Shohara R, Schissel A, Rennie D. 2000. Policies on faculty conflicts of interest at U.S. universities. *Journal of the American Medical Association* 284:2203–2208.

Grinnell F. 1999. Ambiguity, trust, and responsible conduct of research. *Science and Engineering Ethics* 5:205–214.

Gunsalus CK. 1993. Institutional structure to ensure research integrity. *Academic Medicine* 68:S33–S38.

Gunsalus CK. 1998. How to blow the whistle and still have a career afterwards. *Science and Engineering Ethics* 4:51–64.

IOM (Institute of Medicine). 2001. *Preserving Public Trust.* Washington, DC: National Academy Press.

McCabe DL, Pavela GM. 1998. The effect of institutional policies and procedures on academic integrity. In: Burnett DD, Rudolph L, Clifford KO, eds. *Academic Integrity Matters.* Washington, DC: National Association of Student Personnel Administrators, Inc. Pp. 93–108.

NAS (National Academy of Sciences). 1995. *On Being a Scientist,* 2nd ed. Washington, DC: National Academy Press.

NAS. 1997. *Advisor, Teacher, Role Model, Friend: On Being a Mentor to Students in Science and Engineering.* Washington, DC: National Academy Press.

NAS. 2000. *Enhancing the Postdoctoral Experience for Scientists and Engineers.* Washington, DC: National Academy Press.

NIH (National Institutes of Health). 1989. Requirement for programs on the responsible conduct of research in National Research Service Award Institutional Training Programs, p. 1. In: *NIH Guide for Grants and Contracts*, Vol. 18:1, December 22, 1989. Rockville, MD: NIH.

NIH. 2000. *Required Education in the Protection of Human Research Participants NIH Guide for Grants and Contracts*, June 5, 2000 (Revised August 25, 2000). [Online]. Available: http://grants.nih.gov/grants/guide/notice-files/NOT-OD-00-039.html [Accessed December 10, 2001].

NRC (National Research Council). 1996. *Guide for the Care and Use of Laboratory Animals.* Washington, DC: National Academy Press.

Oxford English Dictionary, 2nd ed. 1989. Oxford: Oxford University Press.

Resnik DB. 1998. *The Ethics of Science: An Introduction.* New York: Routledge.

Sigma Xi. 1999. *The Responsible Researcher: Paths and Pitfalls.* Research Triangle Park, NC: Sigma Xi, the Scientific Research Society.

Swazey JP, Anderson MS. 1998. Mentors, advisors, and role models in graduate and professional education. In: Rubin ER, ed. *Mission Management.* Washington, DC: Association of Academic Health Centers. Pp. 165–185.

University of Michigan. 1999. *How to Get the Mentoring You Want: A Guide for Graduate Students at a Diverse University.* [Online] Available: http://www.rackham.umich.edu/StudentInfo/Publications/StudentMentoring/mentoring.pdf [Accessed March 15, 2002].

Yarborough M, Sharp RR. 2002. Restoring and preserving trust in biomedical research. *Academic Medicine* 77:8–14.

3

The Research Environment and Its Impact on Integrity in Research

To provide a scientific basis for describing and defining the research environment and its impact on integrity in research, it is necessary to articulate a conceptual framework that delineates the various components of this environment and the relationships between these factors. In this chapter, the committee proposes such a framework based on an open-systems model, which is often used to describe social organizations and the interrelationships between and among the component parts. This model offers a general framework that can be used to guide the specification of factors both internal and external to the research organization that is relevant to understanding integrity in research.

After its review of the literature, the committee found that there is little empirical research to guide the development of hypotheses regarding the relationships between environmental factors and the responsible conduct of research. Thus, the committee drew on more general theoretical and research literature to inform its discussion. Relevant literature was found in the areas of organizational behavior and processes, ethical cultures and climates, moral development, adult learning and educational practices, and professional socialization.[1]

---

[1]For general references on organizational behavior and processes, see Donabedian (1980), Hamner and Organ (1978), Harrison (1994), Katz (1980), Katz and Kahn (1978), Peters (1978), Peters and Waterman (1982), and Pfeffer (1981). For general references on ethical cultures and climate see Ashforth (1985), Schneider and Reichers (1983), and Victor and Cullen

THE OPEN-SYSTEMS MODEL

The open-systems model depicts the various elements of a social organization; these elements include the external environment, the organizational divisions or departments, the individuals comprising those divisions, and the reciprocal influences between the various organizational elements and the external environment (Ashforth, 1985; Beer, 1980; Daft, 1992; Harrison, 1994; Katz and Kahn, 1978; Schneider and Reichers, 1983). The underlying assumptions of the open-systems model and its various elements are as follows (Harrison, 1994):

1. External conditions influence the inputs into an organization, affect the reception of outputs from an organization's activities, and directly affect an organization's internal operations.

2. All system elements and their subcomponent parts are interrelated and influence one another in a multidirectional fashion (rather than through simple linear relationships).

3. Any element or part of an organization can be viewed as a system in and of itself.

4. There is a feedback loop whereby the system outputs and outcomes are used as system inputs over time, with continual change occurring in the organization.

5. Organizational structure and processes are in part determined by the external environment and are influenced by the dynamics between and among organizational members.

6. An organization's success depends on its ability to adapt to its environment, to tie individual members to their roles and responsibilities within the organization, to conduct its processes, and to manage its operations over time.

THE OPEN-SYSTEMS MODEL OF RESEARCH ORGANIZATIONS

Figure 3-1 shows the application of the open-systems model to the research environment, which can include public and private institutions, such as research universities, medical schools, and independent research organizations. As noted above, any element or part of an organization can

---

(1988). For general references on moral development, see Kohlberg (1984), Rest (1983), and Rest et al. (1999). For general references on adult learning and educational practices, see Brookfield (1986), Cross (1981), and Knowles (1970). For general references on professional socialization, see Schein (1968), Siehl and Martin (1984), Van Maanen and Schein (1979), and Wanous (1980).

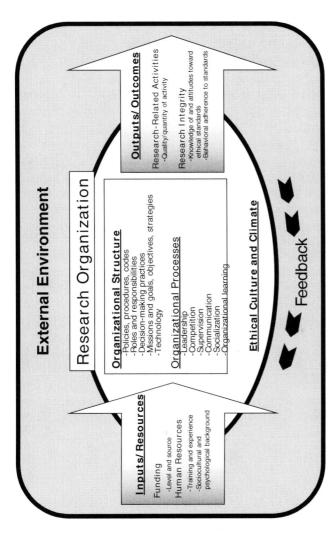

FIGURE 3-1 Open-systems model of the research organization. This model depicts the internal environmental elements of a research organization (white oval), showing the relationships among the inputs that provide resources for organizational functions, the structures and processes that define an organization's operation, and the outputs and outcomes of an organization's activities that are carried out by individual scientists, research groups or teams, and other research-related programs. All of these elements function within the context of an organization's culture and climate. The internal environment is affected by the external environment (shaded area; see Figure 3-2 for further detail). The system is dynamic, and, as indicated by the feedback arrow, outputs and outcomes affect future inputs and resources.

be viewed as a system in and of itself. For research organizations, then, this includes not only the institution itself, but also any of its departments, divisions, research groups, and so on. Figure 3-1 illustrates the research environment as a system that functions within an external environment, whereas Figure 3-2 depicts the specific factors within the external environment and their influence on the research organization. These factors within the external environment are discussed later in this section.

An organization's internal environment consists of a number of key elements—specifically, the *inputs* that provide *resources* for organizational functions, the *organizational structure and processes* that define an organization's setup and operations, and the *outputs and outcomes* that are the results of an organization's activities. The system is dynamic, and, as indicated by the feedback arrow in Figure 3-1, outputs and outcomes affect future inputs and resources. However, all of these components exist within the context of an organization's *culture* and specific *climate* dimensions—that is, the prevailing norms and values that inform individuals within the organization about acceptable and unacceptable behaviors. With respect to the committee's focus on integrity in research, the ethical dimension of the organizational culture and climate is very important.

Structurally, organizations are compartmentalized into various subunits, including work groups or divisions (the research group or team), along with other defined sets of organizational activities and responsibilities (e.g., programs that educate members about the responsible conduct of research, institutional review boards [IRBs], and mechanisms for disclosing and managing conflicts of interest). The operation of these programs and their overall effectiveness influence researchers' perceptions of the organization's ethical climate. Individuals within an organization exist both within and across these defined groups and sets of activities. Given this, it is important to differentiate between an organizational level of analysis (e.g., the research university, medical school, and independent research organization) vis-à-vis the group level of analysis (e.g., the research group or team) and the individual level of analysis (e.g., the individual scientist or researcher).

Inputs and Resources

In its examination of research environments, the committee focused on two input and resource factors of importance: the levels and sources of funding for scientific research, and the characteristics of human resources. These inputs and resources are obtained from an organization's external environment and are used in the production of an organization's outputs.

Funding

The research funding that an organization receives is distributed to research groups or teams and to individual scientists. Funding levels may increase and decrease over the years, both for the organization as a whole and for individual research groups. Just as the overall level of funding available for research within society affects the scientific enterprise as a whole, the level of funding coming into a particular research organization or research team also affects behavior.

The impacts that the level of funding and the competition over funding have on the responsible conduct of research are not clearly understood. There is some limited evidence that in highly competitive environments, individuals with a high "competitive achievement striving" are at risk for engaging in misconduct, particularly when they are faced with situations in which their expectations for success cannot be reached by exerting additional effort (Heitman, 2000; Perry et al., 1990). Encouraging a high level of individual integrity in research, despite vigorous competition for funding, presents a significant challenge for research organizations.

Human Resources

The human resources available to a research organization are also important to the analysis of integrity in research. The background characteristics of scientists coming into a research organization influence its structure and processes as well as its overall culture and climate, and these factors, in turn, influence the responsible conduct of research by individual scientists. Scientists (whether they are trainees, junior researchers, or senior researchers) entering into a research organization will have competing professional demands (e.g., research, teaching, practice, and professional service), and thus there are likely to be conflicting commitments. The dynamics of these competing demands and conflicting commitments change as individual scientists become integrated into the research organization, taking on specific roles and responsibilities.

Also, scientists enter into an organization with various educational and cultural backgrounds. They have different conceptions of the collaborative and competitive roles of the scientist, different abilities to interpret the moral dimensions of problems, and different capacities to reason about and effectively resolve ethical problems. These individual differences will influence organizational behavior, in general, and research conduct, in particular, in complex and dynamic ways.

Given this variation in human resource input into the research organization, it is particularly important for institutions to socialize newcomers and provide them with an understanding of the organization and how to act within it. As in any organization, newcomers must learn the logistics of their organization, the general expectations of their roles by peers, the formal and informal norms governing behavior, the status and power structures, the reward and communication systems, various organizational policies, and so on (Katz, 1980). Within research organizations, individual differences are complicated by the international nature of the scientific workforce and the corresponding sociocultural differences. Therefore, it is particularly important for research institutions to create an environment in which scientists are able to gain an awareness of the responsible conduct of research as it is defined within the culture, to understand the importance of professional norms, to acquire the capacity to resolve ethical dilemmas, and to recognize and be able to address conflicting standards of research conduct.

Organizational Structure and Processes

Structure

To better understand the impact of the research environment on integrity in research, it is important to focus on the organizational elements that characterize its structure—those elements that are more enduring and less prone to change on a day-to-day basis. These elements include an organization's policies and procedures; the roles and responsibilities of members of the organization; decision-making practices; mission, goals and objectives, including the strategies and plans of the organization; and technology.

Policies, Procedures, and Codes   The formalization of policies and practices to support the responsible conduct of research is important in the analysis of research environments and their influence on integrity in research. Chapter 2 identified a number of the practices that are essential to the research environment. Specifically, a research organization should have explicit (versus implicit or nonexistent) procedures and systems in place to fairly (1) monitor and evaluate research performance, (2) distribute the resources needed for research, and (3) reward achievement. These policies and procedures should include criteria related to the responsible conduct of research that are applied consistently. Furthermore, research organizations support integrity in research when they have efficient and effective systems in place to review research involving humans and animals, manage conflicts of interest, respond to misconduct, and socialize trainees and other scientists into responsible research practices. The specification of these policies and procedures helps to regulate and maintain group control and reduce uncertainty about acceptable and unacceptable behaviors (Hamner and Organ, 1978).

Research has shown that strongly implemented and embedded ethical codes of conduct within organizations are associated with ethical behavior in the workplace. McCabe and Pavela (1998) describe the University of Maryland at College Park as one example where implementation of a strong "modified"[2] honor code has proven to be a successful strategy for creating a culture where cheating is viewed as socially unacceptable. Major elements of the Maryland model include (1) involving students in educating their peers and resolving academic dishonesty allegations, (2) treating academic integrity as a moral issue, and (3) promoting enhanced student-faculty contact and better teaching. The mere presence of an honor code, however, is generally not sufficient. Rather, the honor code is used as a vehicle to create a shared understanding and acceptance of the policies on academic integrity among both faculty and students (McCabe and Trevino, 1993).

Corporate codes have a similar effect in the workplace. An original study by McCabe demonstrated that self-reported unethical behavior was lower for survey respondents who worked in a company with a corporate code of conduct (McCabe et al., 1996). Self-reported unethical behavior was inversely correlated with the degree to which the codes were embedded in corporate philosophy and the strength with which the code was implemented (determined by survey questionnaire of employee perceptions).

Roles and Responsibilities The specification of roles and responsibilities within various research groups and teams and relevant research programs (e.g., education in the responsible conduct of research, IRBs, and conflict-of-interest review committees) provides a blueprint for researcher behavior. It is particularly important to clearly define researchers' responsibilities related to the responsible conduct of research. Furthermore, the

---

[2]Traditional honor codes generally include a pledge that students sign attesting to the integrity of their work, a strong, often exclusive role for students in the judicial process that addresses dishonesty allegations, and provisions such as unproctored exams. Some also require students to report any cheating observed. Modified honor codes generally include a strong or exclusive role for students in the academic judicial system, but do not usually require unproctored exams or that students sign a pledge. Modified codes do place a strong campus focus on the issue of academic integrity and students are reminded frequently that their institution places a high value on integrity (McCabe, 2000).

relative positions of these responsibilities within the organizational hier-
archy and the status of persons who operate them will send a clear message to the research community about the importance of such endeavors.
For example, if a highly respected scientist with high status spearheads
the program of education in the responsible conduct of research, and
sufficient resources (in terms of both staff and financial resources) are
available to carry out the program's work, then there is a greater likelihood that its efforts will be taken seriously. Again, these factors have
great symbolic value within the organization and provide compelling
images of the organization's ethical culture, which affects the degree to
which members of the organization will internalize the norms associated
with the responsible conduct of research (Pfeffer, 1981; Siehl and Martin,
1984).

Decision-Making Practices   How an organization reaches decisions and
formulates policies will affect individuals' perceptions of these policies
and their behavioral compliance with them. Individuals are more likely to
accept and adhere to policies and practices when they have played a role
in their development and implementation. Hence, scientists are more
likely to buy into various research policy decisions that are reached
through a collaborative process among key stakeholder groups, rather
than being imposed by a top-level centralized authority (Anderson et al.,
1995, Saraph et al., 1989). Organizational research that focuses on the
pursuit of quality and that explicitly values cooperation and collaboration
to achieve maximum effectiveness leads to better decisions, higher quality, and higher morale within an organization (NIST, 1999). Classically,
faculty and administrators both have governing roles in academic institutions, and this shared responsibility facilitates the bottom-up establishment of rules of research behavior.

Missions, Goals and Objectives, and Strategies and Plans   The mission
and goals of an organization specify its desired end states (e.g., becoming
a "best-practice" site in terms of the protection of human research subjects). Objectives are the specific targets and indicators of goal attainment
(e.g., becoming an accredited program and receiving recognitions and
awards through scientific associations). Strategies and plans are the overall routes and specific courses of action (e.g., allocating the resources to
comply with the standards for accreditation and ensuring that the program has leadership support) to the achievement of goals. If the responsible conduct of research is a prominent part of the mission and goals of a
research organization, along with associated objectives, strategies, and
plans, then the prominence of this issue sets the tone for the organization's

ethical climate and sends a message to scientists that the responsible conduct of research is important. Research has shown that the most successful organizations are those that have a vision and goals that are clearly defined, consistent, and shared among their members (Anderson et al., 1995; Deming, 1986; Freuberg, 1986; Hackman and Wageman, 1995).

Technology   An organization's technology offers the methods for transforming system resources into system outputs. It consists of such aspects of an organization's infrastructure as facilities, tools and equipment, and techniques. These aspects can be mental and social, mechanical, chemical, physical, or electronic. Research environments not only need the necessary tools and equipment for their respective types of scientific research, but they must also establish technologies (e.g., accounting systems and library and information retrieval systems) within the organization for the effective and efficient operation of the research. There may be competition within an organization to acquire the various forms of technology that are of sufficient quantity and quality to facilitate research production. The availability of this technology may, in turn, attract highly skilled scientists who hope to carry out research at the cutting edge of technology. As already mentioned, the effective management of competition—in this case, for technologies—is an important element of promoting the responsible conduct of research.

Processes

Organizational processes, as opposed to an organization's more stable and enduring structural elements, are the patterned forms of interaction between and among groups or individuals within an organization. Processes represent the dynamic aspects of an organization. The processes that characterize organizational dynamics are too numerous to mention here. However, in the committee's examination of research organizations, the processes of most interest consist of (1) leadership, (2) competition, (3) supervision, (4) communication, (5) socialization, and (6) organizational learning.

Leadership   The level of support for high ethical standards by the leadership of an organization or research group can vary; leaders can be extremely supportive, can show ambivalence, or can be nonsupportive. Leaders at every level serve as role models for organizational members and set the tone for an organization's ethical climate (Ashforth, 1985; OGE, 2000; Treviño et al., 1996). Therefore, when leaders support high ethical standards, pay attention to responsible conduct of research, and are openly and strongly committed to integrity in research, they send a clear message about the importance of adhering to responsible research practices (Wimbush and Shepard, 1994). Considerable evidence from the organizational research literature supports the relationship between supervisor behavior and the ethical conduct of the members of an organization (Posner and Schmidt, 1982, 1984; Walker et al., 1979). Supervisors provide a model for how subordinates should act in an organization. Furthermore, supervisors have a primary influence over their subordinates, an influence that is greater than that of an ethics policy. Even if a company or profession has an ethics policy or code of conduct, subordinates follow the leads of their supervisors (Andrews, 1989).

Competition The extent to which the organization is highly competitive, along with the extent to which its rewards (e.g., funding, recognition, access to quality trainees, and power and influence over others) are based on extramural funding and short-term research production, may have negative impacts on integrity in research. Evidence from organizational research indicates that reward systems based on self-interest and commitment only to self rather than to coworkers and the organization are negatively associated with ethical conduct (Kurland, 1996; Treviño et al., 1996). In addition, the level of unethical behavior increases in organizations where there is a high degree of competitiveness among workers (Hegarty and Sims, 1978, 1979). Given these facts, one might expect that a research environment in which competition for resources is fierce and rewards accrue to those who produce the most over the short term sends a wrong message, a message that says "produce at all costs."

Creating a reward system and policies that promote being the "best" within the scientific enterprise, and within a context that encourages the responsible conduct of research, represents a challenge in research environments.

Supervision The extent to which research behavior is monitored and quality control systems are operational will affect the level of adherence to ethical standards. Scientists need to see that policies about responsible research behavior are not just window dressing and that the organization has implemented practices that follow up stated policies. Consistency between words and deeds encourages the members of an organization to take policies seriously. Organizations vary widely in terms of their efforts to communicate codes of conduct to members, as well as to implement mechanisms to ensure compliance. When implementation is forceful and the policies and practices become deeply embedded in an organization's culture, there is a greater likelihood that they will be effective in preventing unethical behavior (McCabe and Treviño, 1993; Treviño, 1990; OGE, 2000).

Communication   Communication among members of a research organization or research group that is frequent and open, versus infrequent and closed, should have a positive influence on integrity in research. A positive ethical climate is supported by open discussions about ethical issues (Jendrek, 1992; OGE, 2000). Frequent and open communication enhances awareness of issues, encourages individuals to seek advice when faced with ethical dilemmas, and establishes the importance of resolving issues before they become something to be hidden.

Socialization   Mentoring relationships between research trainees and their advisers are important in the socialization of young scientists (Anderson et al., 2001; Swazey and Anderson, 1998). These relationships can be characterized by a variety of factors, including the level of trust, communication patterns, and the fulfillment of responsibilities as a mentor or trainee. In addition to mentoring relationships, education in research and professional ethics is an aspect of socialization (Anderson, 1996; Anderson and Louis, 1994; Anderson et al., 1994; Louis et al., 1995; Swazey et al., 1993). Socialization practices can be formal or informal, but they are essential to helping individuals internalize the norms and values associated with the responsible conduct of research. Research that examines the effect of more formalized methods of socialization—for example, education—reveals that interactive techniques (e.g., case discussion, role-playing, and hands-on practice sessions) are generally more effective in producing behavioral change than are activities with minimal participant interaction or discussion (e.g., lectures or presentations [Davis et al., 1999]). Furthermore, sequenced education has a greater impact than single educational sessions (Davis et al., 1999; OGE, 2000). These findings substantiate the principles of adult education; these principles describe successful practices as being learner-centered, active rather than passive, relevant to the learner's needs, engaging, and reinforcing (Brookfield, 1986; Cross, 1981; Knowles, 1970) (Chapter 5).

Organizational Learning   Organizations that learn from their operations and that continuously seek to improve their performance are better able to adapt to a changing environment (Anderson et al., 1994; Deming, 1986; Hackman and Wageman, 1995; Schön, 1983). All organizations change over time, but for some this can be an excruciating and painful process if it comes about through reaction to a crisis situation. For example, when a research subject dies or a researcher is accused of data fabrication, the organization should respond immediately. However, this response is focused on crisis intervention rather than prevention. On the other hand, organizations that have mechanisms in place to continuously evaluate the efficiency and effectiveness of their programs and activities are more likely to build a preventive maintenance system (Fiol and Lyles, 1985; Schön, 1983). Furthermore, if the members of an organization have a voice in the design and implementation of such systems, then they are more likely to accept and be cooperative with the continual evaluative processes.

Culture and Climate

All of the enduring elements and features of an organization's structure and its more dynamic processes exist within the context of an organization's culture and climate. In fact, an organization's structure and processes help to create the culture and climate inasmuch as they are shaped by them (Ashforth, 1985). An organization's culture consists of the set of shared norms, values, beliefs, and assumptions, along with the behavior and other artifacts (e.g., symbols, rituals, stories, and language) that express these orientations.. Culture and climate factors are characteristics of an organization that guide members' thoughts and actions (Schneider, 1975).

The ethical (or moral) climate is one component of an organization's culture and is particularly relevant in the analysis of integrity in research (Victor and Cullen, 1988). This climate is defined as the prevailing moral beliefs (i.e., the prescribed behaviors, beliefs, and attitudes within the community and the sanctions expressed) that provide the context for conduct. The stable, psychologically meaningful, and shared perceptions of the members of an organization are used as indicators of ethical climate, which may exist both at the organizational level and at the research group or team level (Schneider, 1975; Schneider and Reichers, 1983).

An ethical climate that supports the responsible conduct of research is created when scientists perceive that adherence to ethical standards takes precedence and that sanctions for ethical violation are consistently applied. Research in this area has established that the factors within an organization that are most strongly related to ethical behavior are attention to ethics by supervisors and organizational leadership, consistency between policies and practices, open discussions about ethics, and follow-up of reports of ethics concerns (OGE, 2000). These features of an organization can help establish an ethical climate in which organizational members perceive that the responsible conduct of research is central to the organization's practice and that it is not something to be worked around. It creates an environment in which a code of conduct is strongly implemented and deeply embedded in the community's culture (Treviño, 1990).

Outputs and Outcomes

Outputs

The outputs of research organizations are produced at all levels—the organizational level, the research group or team level, and the individual scientist level. The outputs are the products produced, the services delivered, and the ideas developed and tested. The most obvious outputs are the number and quality of research projects completed, reports written, publications produced, patents filed, and students graduated.

For the committee's purposes, however, it is important to focus on the outputs of activities or programs related to integrity in research—for example, institutional review boards, conflict-of-interest review committees, and programs that provide education in the responsible conduct of research. Outputs from these programs are generally measured in terms of the quantity and the quality of activities—for example, the number of workshops and seminars offered, the number of scientists who participate, and the number of research proposals reviewed by IRBs and the dispositions of those proposals. Research organizations that design and implement high-quality activities related to integrity in research—and in a quantity that is sufficient to meet their needs—are more likely to achieve the outcomes that they seek (e.g., adherence to responsible research practices). Although these activities will not be the sole factors that determine the responsible conduct of research, their implementation becomes a symbol for the members of an organization, serving as an indicator of the leadership's commitment to the establishment of a culture and a climate that supports the responsible conduct of research.

Outcomes

The outcomes of organizational activities refer to the specific results that reflect the achievement of goals and objectives. As with organizational outputs, outcomes can be associated with the organization as a whole, the research group, or the individual scientist. However, the committee's primary interest is in the individual scientist's level of integrity in research. As discussed in Chapter 2, the committee defines integrity in research as the individual scientist's adherence to a number of normative practices for the responsible conduct of research.

Adherence to these practices provides a set of behavioral indicators of an individual's integrity in research. However, behavioral compliance is assumed to be associated with an understanding of the norms, rules, and practices of science. In addition, judgments about an individual's integrity are based on the extent to which intellectual honesty, accuracy, fairness, and collegiality consistently characterize the dispositions and attitudes reflected in a researcher's practice. Judgments about a person's integrity are less about strict adherence to the rules of practice and are more about the disposition to be intellectually honest, accurate, and fair in the practice of science (i.e., in the willingness to admit and correct one's errors and shortcomings).

The committee resisted defining integrity in terms of (1) adherence to the normative practices listed in Chapter 2, (2) the knowledge and awareness of the practices of responsible research, and (3) the attitudes and orientation toward the practices of responsible research (i.e., the degree to which individuals agree with the practices, the level of importance that they attach to them, and the extent to which they are subject to conflicting sets of practices), as has been common in the social sciences.[3] These three conceptually distinct categories of outcomes fail to capture the complexity of the process through which individuals interact with their environment and make ethical decisions. One simply cannot assume that as scientists gain awareness of standards of practice, they will be positively oriented to them or will be more likely to adhere to the behavioral requirements. The committee recognizes that although researchers might be well intentioned, there is truth in what psychologists (Rest, 1983) have observed: that everyone is capable of missing a moral issue (moral blindness); developing elaborate and internally persuasive arguments to justify questionable actions (defective reasoning); failing to prioritize a moral value over a personal one (lack of motivation or commitment); being ineffectual, devious, or careless (character or personality defects, often implied when someone is referred to as "a jerk"); or having ineffectual skills at problem solving or interpersonal communication (incompetence).

For this reason, focusing on the processes that give rise to the responsible conduct of research are important individual-level outcomes of organizational activities within the research environment. Components of the process of ethical decision making include ethical sensitivity, reason-

---

[3]A recent review of approaches to the study of morality (Bebeau et al., 1999) has challenged the usefulness of the usual tripartite view that assumes that the elements to be studied and assessed are attitudes, knowledge, and behavior. When researchers have studied the connections among these elements, they usually do not find significant connections and are left with the conclusion that attitudes do not have much to do with knowing and behavior is often devoid of feeling and thinking. A more profitable approach is to assume that many types of cognitions, many types of affects, and many kinds of observable behaviors are involved in morality or integrity. All behavior is the result of cognitive-affective processes. Instead of studying cognitions, affects, and behaviors as separate elements, psychologists suggest that researchers study functional processes that must arise to produce moral behavior (Rest, 1983).

ing, moral motivation and commitment, and character and competence (Bebeau, 2001). Educational programs that train scientists in the responsible conduct of research are often premised on the assumption that these essential capacities for ethical decision making are well developed by the time individuals begin their research education, and that one simply needs to teach the rules of the responsible conduct of research. Research on ethical development in the professions demonstrates that even mature professionals show considerable variability on performance assessments that measure ethical sensitivity, moral reasoning and judgment, professional role orientation, and appropriate character and competence to implement action plans effectively.

Therefore, if a research environment implements educational programs to foster integrity in research, then these programs should promote sensitivity to issues that are likely to arise in the research setting by building a capacity for reasoning carefully about conflicts inherent in proposing, conducting, and reporting research; by developing a sense of personal identity that incorporates the norms and values of the research culture; and by building competence in problem solving and interpersonal communication (see Chapter 5 for further discussion).

External Environment

The external environment of a research organization consists of both an external-task environment and a general environment (Figure 3-2). The external-task environment includes all the organizations and conditions that are directly related to an organization's main operations and its technologies. The systems and subsystems of the external-task environment are embedded within the larger sociocultural, political, and economic environment and have a more indirect impact on an organization. It is important to recognize that relationships also exist between and among all elements within the external environment. For example, government policies and regulations can affect the areas and levels of funding. Journal policies can be affected by decisions made within scientific associations, and these decisions can be driven by government regulation (or pending regulation).

External-Task Environment

A number of factors within the external-task environment have a significant impact on scientists' responsible conduct of research. These factors include government regulation, funding for scientific work, job opportunities for trainees and researchers, journal policies and practices, and the policies and practices of scientific societies.

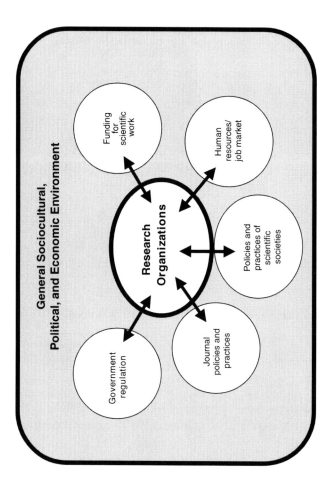

FIGURE 3-2 Environmental influences on integrity in research that are external to research organizations. The *external-task environment* includes all of the organizations and conditions that are directly related to an organization's main operations and technologies. The double arrows depict the interrelatedness between the research organization and the various external influences (unshaded circles) that are hypothesized to have an impact on integrity in research. The *general environment* has a more indirect impact on an organization. The systems and subsystems of the external-task environment are embedded within the larger, general sociocultural, political, and economic environment (shaded area). Although not specifically shown in this figure, it is important to recognize that relationships exist between and among the elements within the external environment.

Government Regulation Governmental bodies, particularly at the federal level, have been promulgating regulations concerning the conduct of research for many years. Most widely known and recognized are the regulations regarding the protection of human research subjects (45 C.F.R. § 46, 1999; 21 C.F.R. § 50 and 56, 1998) and the protection of animals in research (7 U.S.C. §§ 2131, 1966, et seq.). Furthermore, regulations have been promulgated regarding the evaluation of allegations and the reporting of scientific misconduct (42 C.F.R. § 50, §§A, 1989; *Federal Register*, 2000) and the handling and disposal of hazardous chemicals in the laboratory (29 C.F.R. § 1910.1450, 1996), to name just two. As these government regulations come into force, they have direct impacts on a research organization and individual scientists. Specifically, organizations and individuals must be in compliance with the regulations or face sanctions.

Funding for Scientific Work Research organizations are directly affected by both the level and the source of funding that is available for scientific work (e.g., they are affected by the balances between government and corporate support and between industry and foundation support). Most funding sources provide support for specific research proposals rather than particular investigators. Although proposals are usually ranked on a relative scale, more typically they are funded in an all-or-none fashion. At the same time, funding needs always outpace funding opportunities. For instance, only one in three investigator-initiated grant proposals (see http://silk.nih.gov/public/cbz2zoz.@www.com.rpg.act. dsncc) to the National Institutes of Health is successful. In this situation, even investigators who succeed in their research sometimes lose funding, a fate that threatens the very existence of their projects and often threatens their personal incomes.

The task for research organizations is to develop structures that help their scientists deal with this competitive research situation while maintaining the responsible conduct of research. Similarly, when corporate or industry funds are involved, research organizations should require strategies for the management and disclosure of conflicts of interest to reduce problems related to publication rights, ownership of intellectual property, and research involving human subjects.

Job Opportunities When the job market is tight and there is more competition for every research position, researchers will be pressured to achieve higher levels of productivity and recognition. This situation challenges scientists to be the best while maintaining the highest levels of integrity in research. Similarly, research programs must compete for students and postdoctoral fellows, who, in turn, enhance a program's accomplishments and overall status. The ability of researchers to gain recognition often is believed to be the best path to attracting high-quality trainees to a program. The organizational challenge is to help researchers develop competitive programs while maintaining a high level of commitment to integrity in research.

Journal Policies and Practices   Journal editors can be more or less rigorous in their implementation of the review process and the extent to which they insist on high levels of adherence to scientific standards. Furthermore, journals may have specific policies in such areas as authorship practices, disclosure of conflicts of interest, duplicate publication, and reporting of research methodologies. The scientific community receives an important message about integrity in research when journal policies and practices regarding these practices are clear and are required as a condition of publication—and when the most prestigious journals adopt such practices. For example, members of the International Committee of Medical Journal Editors recently revised their submission policies related to industry-sponsored research. Authors are now required to sign a statement accepting full responsibility for the conduct of a clinical trial, and they must confirm that they had access to the original data and had full control over the decision to publish (Davidoff et al., 2001).

Policies and Practices of Scientific Societies   Scientific societies are in a position to influence the behaviors of their members in ways that could promote integrity in research[4] (AAAS, 2000). The societies vary extensively, however, in their development of codes of conduct, their enforcement of such codes, and their socialization of members with regard to these standards of behavior. To aid in this process, the Association of American Medical Colleges has published a guide to help societies in the development of ethical codes (AAMC, 1997). Other associations develop standards for accreditation—for example, standards for science education programs, research laboratories, and programs for the protection of human and animal research subjects. These accreditation standards generally have specific statements regarding the responsible conduct of research and stipulate the structures within the organization that must be in place to ensure compliance with the standards. Scientists who are part of such accredited programs will be subject to the influences of these external controls.

---

[4]See Chapter 6 for further discussion of the role professional and scientific societies can play in fostering an environment that promotes integrity in research.

General Environment

The general environment has an indirect impact on an organization. This environment includes all of the conditions and institutions that have sustained or infrequent impacts on the organization and its functions (Harrison, 1994). Included are the state or conditions of major social institutions (e.g., the economy, political system, educational system, science and technology system, and legal system) as well as the local, national, and international cultures within which an organization operates. The general public, and more specifically the effects of public trust in the research enterprise, are also important components of the general environment. As reflected in Figure 3-2, the organizations and conditions of the external-task environment (unshaded circles) are embedded within this larger environment (shaded area).

An example of how the broader environment can affect the conduct of research is the recent national debate over embryonic stem cell research; this debate reflects a clash of values that affect the characterization of ethical or unethical research (NAS, 2001; National Bioethics Advisory Commission, 1999). In another instance, the new rules governing the privacy of health records that are part of the Health Insurance Portability and Accountability Act are being challenged by scientists as too restrictive in providing access to identifiable data for research (AAMC, 2001; Annas, 2002). Also, society places a high premium on human rights and the protection of vulnerable persons, values that have been translated into federal regulations for the protection of human research subjects (45 C.F.R. § 46, 1993, and 21 C.F.R. § 50 and 56, 1981).

Other social institutions also have an indirect impact on research environments. Educational systems produce scientists, and these systems affect not only their quantity but also their quality and how well they have been socialized into professional standards of conduct. The technology systems determine the availability of equipment and the methods used to carry out various types of research, factors that may raise questions about the propriety of certain research endeavors. Ethical conflicts are often created when the development of new technologies requires an answer to the question of whether what can be done should be done. Finally, the legal system and the propensity in the United States to resort to litigation may bring about situations in which scientists are caught between the responsible conduct of research and subpoenas for confidential data. These examples are by no means exhaustive, but they reflect the ways in which major social institutions and cultural values can affect research organizations and a scientist's practice of research.

SUMMARY

The committee found no comprehensive body of research or writing that can guide the development of hypotheses regarding the relationships between the research environment and the responsible conduct of research. However, viewing the research environment as an open-systems model, which is often used in general organizational and administrative theory, makes it possible to hypothesize how various components affect integrity in research. Inputs of funds and other resources can influence behavior both positively and negatively. The organizational structure and processes that typify the mission and activities of an organization can either promote or detract from the responsible conduct of research. The culture and climate that are unique to an organization both promote and perpetuate certain behaviors. Finally, the external environment, over which individuals and, often, institutions have little control, can affect behavior and alter institutional integrity for better or for worse.

REFERENCES

AAAS (American Association for the Advancement of Science). 2000. *The Role and Activities of Scientific Societies in Promoting Research Integrity.* A report of a conference, April 10, 2000, Washington, DC. [Online]. Available: http://www.aaas.org/spp/dspp/sfrl/projects/integrity.htm [Accessed January 7, 2002].

AAMC (Association of American Medical Colleges). 1997. *Developing a Code of Ethics in Research: A Guide for Scientific Societies.* Washington, DC: AAMC.

AAMC. 2001. *Letter to Secretary Thompson on Impact of Medical Privacy Rule on Research.* [Online]. Available: http://www.aamc.org/advocacy/corres/research/thompson. htm [Accessed February 1, 2002].

Anderson J, Rungtusanatham M, Schroeder R, Devaraj S. 1995. A path analytic model of a theory of quality management underlying the Deming management method: Preliminary empirical findings. *Decision Sciences* 26:637–658.

Anderson MS. 1996. Misconduct and departmental context: Evidence from the Acadia Institute's Graduate Education Project. *Journal of Information Ethics* 5(1):15–33.

Anderson MS, Louis KS. 1994. The graduate student experience and subscription to the norms of science. *Research in Higher Education* 35:273–299.

Anderson MS, Louis KS, Earle J. 1994. Disciplinary and departmental effects on observations of faculty and graduate student misconduct. *Journal of Higher Education* 65:331–350.

Anderson MS, Oju EC, Falkner TMR. 2001. Help from faculty: Findings from the Acadia Institute Graduate Education Study. *Science and Engineering Ethics* 7:487–503.

Andrews KR. 1989. Ethics in practice. *Harvard Business Review* Sept-Oct:99–104.

Annas GJ. 2002. Medical privacy and medical research—judging the new federal regulations. *New England Journal of Medicine* 346:216–220.

Ashforth BE. 1985. Climate formation: Issues and extensions. *Academy of Management Review* 10:837–847.

Bebeau MJ. 2001. Influencing the moral dimensions of professional practice: Implications for teaching and assessing for research integrity. In: Steneck NH, Scheetz MD, eds. [2000]. *Investigating Research Integrity: Proceedings of the First ORI Research Conference on*

Research Integrity. Washington, DC: Office of Research Integrity, U.S. Department of Health and Human Services. Pp. 179-188.

Bebeau MJ, Rest JR, Narvaez D. 1999. Beyond the promise: A perspective on research in moral education. *Educational Researcher* 28(4):18–26.

Beer M. 1980. *Organizational Change and Development—A Systems View.* Santa Monica, CA: Goodyear.

Brookfield SD. 1986. *Understanding and Facilitating Adult Learning: A Comprehensive Analysis of Principles and Effective Practices.* San Francisco, CA: Jossey-Bass.

Cross KP. 1981. *Adults as Learners: Increasing Participation and Facilitating Learning.* San Francisco, CA: Jossey-Bass.

Daft R. 1992. *Organizations: Theory and Design,* 4th ed. St. Paul, MN: West Publishing.

Davidoff F, DeAngelis CD, Drazen JM, Nicholls MG, Hoey J, Hojgaard L, Horton R, Kotzin S, Nylenna M, Overbeke AJPM, Sox HC, Van Der Weyden MB, Wilkes MS. 2001. Sponsorship, authorship and accountability. *Canadian Medical Association Journal* 165:786–788.

Davis D, O'Brien MAT, Freemantle N, Wolf FM, Mazmanian P, Taylor-Vaisey A. 1999. Impact of formal continuing medical education: Do conferences, workshops, rounds, and other traditional continuing education activities change physician behavior or health care outcomes? *Journal of American Medical Association* 282:867–874.

Deming WE. 1986. *Out of Crisis.* Cambridge, MA: Center for Advanced Engineering Study, Massachusetts Institute of Technology.

Donabedian A. 1980. *Explorations in Quality Assessment and Monitoring: The Definition of Quality and Approaches to Its Assessment.* Vol. 1. Ann Arbor, MI: Health Administration Press.

Federal Register. 2000. *Public Health Service Standards for the Protection of Research.* [Online]. Available: http://frwebgate4.access.gpo.gov/cgi-bin/waisgate.cgi?WAISdocID= 47426132664+0+2+0&WAISaction=retrieve [Accessed March 18, 2002].

Fiol DC, Lyles MA. 1985. Organizational learning. *Academy of Management Review* 10:803–813.

Freuberg D. 1986. *The Corporate Conscience: Money, Power, and Responsible Business.* New York, NY: American Management Association.

Hackman JR, Wageman R. 1995. Total quality management: Empirical, conceptual, and practical issues. *Administrative Science Quarterly* 40:309–342.

Hamner WC, Organ DW. 1978. *Organizational Behavior: An Applied Psychological Approach.* Dallas, TX: Business Publications.

Harrison MI. 1994. *Diagnosing Organizations: Methods, Models, and Processes,* 2nd ed. Thousand Oaks, CA: Sage.

Hegarty WH, Sims HP. 1978. Some determinants of unethical decision behavior: An experiment. *Journal of Applied Psychology* 63:451–457.

Hegarty WH, Sims HP. 1979. Organizational philosophy, policies, and objectives related to unethical decision behavior: A laboratory experiment. *Journal of Applied Psychology* 64:331–338.

Heitman E. 2000. Ethical values in the education of biomedical researchers. *Hastings Center Report* 30:S40–S44.

Jendrek MP. 1992. Students' reactions to academic dishonesty. *Journal of College Student Development* 33:260–273.

Katz D, Kahn R. 1978. *The Social Psychology of Organizations,* 2nd ed. New York, NY: Wiley.

Katz R. 1980. Time and work: Toward an integrative perspective. In: Staw BM, Cummings LL, eds. *Research in Organizational Behavior,* Vol. 2. Greenwich, CT: JAI Press. Pp. 81–127.

Knowles MS. 1970. *The Modern Practice of Adult Education: Andragogy Versus Pedagogy.* New York, NY: New York Association Press.

Kohlberg L. 1984. *The Psychology of Moral Development: The Nature and Validity of Moral Stages.* San Francisco: Harper & Row. Essays on Moral Development Vol. 2.

Kurland N. 1996. Trust, accountability, and sales agents' dueling loyalties. *Business Ethics Quarterly* 6:289–310.

Louis KS, Anderson MS, Rosenberg L. 1995. Academic misconduct and values: The department's influence. *The Review of Higher Education* 18:393–422.

McCabe DL. 2000. New research on academic integrity: the success of "modified" honor codes. *Synfax Weekly Report* [Online]. Available: http://www.collegepubs.com/ref/SFX000515.shtml [Accessed July 25, 2001].

McCabe DL, Pavela GM. 1998. The effect of institutional policies and procedures on academic integrity. In: Burnett DD, Rudolph L, Clifford KO, eds. *Academic Integrity Matters.* Washington, DC: National Association of Student Personnel Administrators, Inc. Pp. 93–108.

McCabe DL, Treviño LK. 1993. Academic dishonesty: Honor codes and other contextual influences. *Journal of Higher Education* 64:522–538.

McCabe DL, Treviño LK, Butterfield KD. 1996. The influence of collegiate and corporate codes of conduct on ethics-related behavior in the workplace. *Business Ethics Quarterly* 6:461–476.

NAS (National Academy of Sciences). 2001. *Biological and Biomedical Applications of Stem Cell Research.* Washington, DC: National Academy Press.

National Bioethics Advisory Commission. 1999. *Ethical Issues in Human Stem Cell Research,* Vol. I to III. Washington, DC: U.S. Government Printing Office.

NIST (National Institute of Standards and Technology). 1999. *Malcolm Baldrige National Quality Award 1998 Education Criteria for Performance Excellence.* Washington, DC: U.S. Department of Commerce.

OGE (U.S. Office of Government Ethics). 2000. *Executive Branch Employee Ethics: Survey 2000.* Washington, DC: OGE.

Perry A, Kane K, Bernesser K, Spicker P. 1990. Type A behavior, competitive achievement-striving, and cheating among college students. *Psychological Reports* 66:459–465.

Peters TJ. 1978. Symbols, patterns, and settings: An optimistic case for getting things done. *Organizational Dynamics* 7(2):3–23.

Peters TJ, Waterman RH, Jr. 1982. *In Search of Excellence: Lessons from America's Best-Run Companies.* New York, NY: Harper & Row.

Pfeffer J. 1981. Management as symbolic action: The creation and maintenance of organizational paradigms. In: Cummings LL, Staw BM eds. *Research in Organizational Behavior,* Vol. 3. Greenwich, CT: JAI Press. Pp. 1–52.

Posner B, Schmidt W. 1982. What kind of people enter the public and private sectors? An undated comparison of perceptions, stereotypes, and values. *Human Resource Management* 21:35–43.

Posner B, Schmidt W. 1984. Values and the American manager: An update. *California Management Review* 26(3):202–216.

Rest, J. 1983. Morality. In: Mussen PH (series ed.) and Flavell J, Markman E (vol. eds.), *Handbook of Child Psychology,* Vol. 3, *Cognitive Development,* 4th ed. New York, NY: Wiley. Pp. 556–629.

Rest J, Narvaez D, Bebeau MJ, Thoma SJ. 1999. *Postconventional Moral Thinking: A Neo-Kohlbergian Approach.* Hillsdale, NJ: L. Erlbaum Associates.

Saraph JV, Benson PG, Schroeder RG. 1989. An instrument for measuring the critical factors of quality management. *Decision Sciences* 20:810–829.

Schein EH. 1968. Organizational socialization and the profession of management. *Industrial Management Review* 9:1-15.

Schneider B. 1975. Organizational climates: An essay. *Personnel Psychology* 28:447–479.
Schneider B, Reichers AE. 1983. On the etiology of climates. *Personnel Psychology* 36:19–39.
Schön DA. 1983. Organizational learning. In: Morgan G, ed. *Beyond Method: Strategies for Social Research.* Newbury Park, CA: Sage. Pp. 114–128.
Siehl C, Martin J. 1984. The role of symbolic management: How can managers effectively transmit organizational culture? In: Hunt JG, Hosking DM, Schriesheim CA, Stewart R, eds. *Leaders and Managers: International Perspectives on Managerial Behavior and Leadership.* Elmsford, NY: Pergamon. Pp. 227–239.
Swazey JP, Anderson MS. 1998. Mentors, advisors, and role models in graduate and professional education. In: Rubin ER, ed. *Mission Management: A New Synthesis,* Vol. 2. Washington, DC: Association of Academic Health Centers. Pp. 165–185.
Swazey JP, Anderson MS, Louis KS. 1993. Ethical problems in academic research. *American Scientist* 81:542–553.
Treviño LK. 1990. A cultural perspective on changing and developing organizational ethics. *Research in Organizational Change and Development* 4:195–230.
Treviño LK, Butterfield KD, McCabe DL. 1996. The ethical context in organizations: Influences on employee attitudes and behaviors. *Business Ethics Quarterly* 8(3):447–476.
Van Maanen J, Schein EH. 1979. Toward a theory of organizational socialization. In: Staw BM, ed. *Research in Organizational Behavior,* Vol. 1. Greenwich, CT: JAI Press. Pp. 209–264.
Victor B, Cullen JB. 1988. The organizational bases of ethical work climates. *Administrative Science Quarterly* 33:101–125.
Walker OC, Churchill GA, Ford NM. 1979. Where do we go from here? Selected conceptual and empirical issues concerning the motivation and performance of the industrial sales force. In: Albuam G, Churchill GA, eds. *Critical Issues in Sales Management State of the Art and Future Research Needs.* Eugene, OR: University of Oregon Press. Pp. 10–75.
Wanous JP. 1980. *Organizational Entry: Recruitment, Selection, and Socialization of Newcomers.* Reading, MA: Addison-Wesley.
Wimbush JC, Shepard JM. 1994. Toward an understanding of ethical climate: Its relationship to ethical behavior and supervisory influence. *Journal of Business Ethics* 13:637–647.

4

Institutional Approaches to Fostering Integrity in Research

Research organizations currently rely on a variety of methods for promoting integrity in research. They establish organizational components to comply with regulations imposed by an external environment; they offer educational programs to teach the elements of the responsible conduct of research;[1] they implement policies and procedures that delineate the normative practices of responsible research and establish criteria for rewards and recognition; and they develop processes to evaluate and enforce institutional behavior.[2] In addition, organizations engage in activities that help establish an internal climate and organizational culture that are either supportive of or ambivalent toward the responsible conduct of research.[3] These various approaches are not mutually exclusive, however, nor should they be. A number of programs and activities, integrated across the various levels of an organization, should be in place in order to maximize the impact on the research environment and to support the responsible conduct of research.

---

[1]Chapter 5 provides a detailed discussion of promoting integrity in research through education.

[2]Chapters 2 and 3 provide discussions of the types of policies and procedures that should be implemented in a research organization to foster integrity in research.

[3]Chapter 3 provides a discussion of the elements of an organization's structure and processes that have an impact on how members of an organization perceive the ethical climate and culture.

To establish a basis for organizational learning and continuous quality improvement, organizations should simultaneously implement processes for evaluating their efforts to foster responsible conduct of research. Evaluation can be approached in a variety of ways. One way is to rely on external evaluators to determine compliance with regulatory controls. Another is to rely on a system of performance-based assessments that are initiated and implemented internally. Such assessments can also be used to meet the accountability requirements of outside funding or government sources. In addition, peer reviewers may be used in institutional self-assessment processes; assessments done by peer reviewers may or may not be associated with accreditation by external organizations.

This chapter provides a discussion of the strengths and limitations of these various approaches and establishes a rationale for the use of the institutional self-assessment approach to evaluation that is recommended in this report (see Chapter 6).

REGULATORY COMPLIANCE

The U.S. Congress has consistently affirmed the importance of integrity in research, and the federal government has established a framework for regulating misconduct in science. As described in Chapter 1, regulatory language includes a definition of research misconduct and spells out the requirements for institutional policies and practices to handle reports of misconduct or other types of wrongdoing in the research environment.

A regulatory framework requires a rule-making process that may be governed by legislative or administrative actions. The framework may be decentralized (i.e., each institution may develop its own approach), or it may require compliance with a common set of criteria, policies, and practices. The safety standards promulgated by the Occupational Safety and Health Administration (OSHA) are an example of the latter. Regulatory models commonly involve data collection and the generation of reports that document each institution's compliance with governmental policies. OSHA, for example, very clearly specifies requirements for recording and reporting on-the-job injuries and illnesses, and is authorized to conduct workplace inspections to check reporting and insure compliance.

Assessment Strategy

Common components of regulatory frameworks include the specification of certain procedures and reporting requirements, the collection of data, and the preparation of reports of compliance practices. The regulatory approach also involves a governmental unit that maintains oversight of the compliance and reporting procedures, investigates complaints about rule violations, and offers technical assistance in rule-making and implementation of regulations. In university settings, evaluations for mandatory compliance with government regulatory standards are common. In this framework, full-time government employees frequently (but not always) conduct inspections to evaluate compliance with a well-defined set of standards, followed by some mechanism for feedback (usually an exit interview or a written report). Individual inspectors do not necessarily require professional credentials in the field or activity being evaluated, although many individuals in such positions have such qualifications (i.e., they are not necessarily "peers" of the individuals whose conduct is being evaluated). In some highly regulated research environments, such as those in which new drugs are developed, iterative partnerships can evolve between regulators (e.g., the Food and Drug Administration) and the regulated entities (e.g., pharmaceutical companies).

Strengths

Several models of regulatory frameworks for research that could be adapted to the oversight of integrity in research already exist. These models include the regulatory frameworks for the oversight of the protection of human research subjects (The Common Rule), the evaluation of misconduct in science, the use of animals in research, and the handling of toxic or radioactive research materials.

A regulatory approach to fostering integrity in research is consistent with other governmental efforts to encourage the use of commonly accepted practices and to discourage irresponsible behavior in the research environment. Researchers and institutional officials are familiar with compliance requirements and often participate in the preparation of rule-making procedures. Individual research centers frequently have some latitude and discretion in adapting government requirements to their own needs, and the centers are responsible for designating specific officials who will ensure that faculty, trainees, and administrative staff understand the importance of the regulations. A regulatory framework fosters a collective and consistent response to social concerns by a broad array of research institutions, and it highlights best practices. The framework also can reduce the opportunity for idiosyncratic and irresponsible policies or practices to be implemented.

Limitations

A regulatory approach fostering integrity in research also has some limitations (*Nature*, 2001; Burman, et al., 2001; Sugar, 2002). Such an approach increases the bureaucratization of science and requires documentation that institutions may find burdensome. A 1999 report commissioned by the National Institutes of Health (NIH) at the request of Congress reviewed the burden of regulations across five key areas, including research integrity. The report identified four problems that were present in all the areas: rigid regulations that dictate process often limit flexibility without enhancing results; rules imposed by multiple agencies have inconsistent requirements; regulation of science by nonscience agencies often leads to additional and nonproductive regulatory burden; and poor communication exists among federal agencies and research institutions (NIH, 1999). The report also notes that regulatory systems tend to impose the same requirements on all research institutions regardless of their characteristics. That is, requirements are aimed at the lowest performer, and hence provide little incentive for superior performance.

Regulations often emphasize the areas of common agreement and can reduce important concerns to rules and procedures. It is difficult or impossible for regulations alone to foster an understanding of the critical issues involved, and the required procedures are not always related to the desired outcomes. The adoption of new regulations and the creation of institutional and governmental oversight offices increase the cost of doing science and add to the administrative costs of research centers without necessarily creating a commensurate benefit. In addition, once regulations are adopted, they are difficult to change. Finally, in some instances regulators are forced to focus more on process than on results. That is, the regulators seek to evaluate the extent to which observable processes that are hypothesized to promote some desired outcome are in place; examples of such processes might include protocols to protect human or animal research subjects, measures to ensure laboratory safety, or desirable data-handling practices. Although a focus on process can facilitate intended outcomes, it can also be directed to matters that are relatively unimportant or that are poorly connected with the desired result.

A PERFORMANCE-BASED APPROACH

A performance-based model for the evaluation of organizational efforts to foster integrity in the research environment offers selected goals and benchmarks[4] that can be used as criteria to assess the success of

---

[4]A benchmark is a standard or point of reference used in measuring and/or judging quality or value. Benchmarking is the process of continuously comparing and measuring an organization's performance, practices, policies, and philosophies against leading, high-performing organizations anywhere in the world to gain information that will help the organization take action to improve its performance (NPR, 1997).

efforts. These goals and benchmarks are generally linked to rewards, incentives, and, at times, penalties for specific types of behavior. A performance-based approach provides a direct role for research institutions and research team leaders in fostering norms for faculty, research staff, trainees, and students within diverse research settings.

The development of a performance-based model would require institutions to formulate a coherent statement of goals that describes the principles of integrity in research that they wish to encourage. The model would also require institutions to implement these goals through a series of actions and assessment strategies. Such actions could include the following:

• posting the statement (including selected criteria related to personnel actions, such as recruitment offers and hiring and promotion policies and practices) in public places throughout the research institution;
• creating a bonus plan or award system to reward exceptional behavior;
• providing mentorship opportunities for senior and junior faculty and investigators that emphasize the importance of learning about the responsible conduct of research; and
• publicizing and possibly sanctioning actions that are inconsistent with the institution's research mission.

Performance-based systems require that benchmarks be set and that a database be generated in order to measure faculty and trainee compliance with the specified standards to identify areas that need improvement. Such benchmarks can be formal or informal, but they require a broad consensus that what is being measured is important and relevant to integrity in research. Performance benchmarks must be communicated clearly so that all members of a research institution understand what is being measured and why. Performance-based approaches should be implemented incrementally, because data must be collected and analyzed before major institutional changes can be implemented. Such implementation is in keeping with the spirit of continuous quality improvement.

Research on organizational behavior found that institutions have ethical climates that differ according to the values, standards, and interests of their members (Victor, 2001). Performance incentives are tools that institutions use to align their members' behaviors and practices with the values that are expressed as the institutions' missions and goals. These tools represent one part of a broader set of normative control systems for organizations, and they are shaped not only by aspirational goals but also by societal norms, the form of the organization, the behavior of the leadership, and unique features of research groups.

Assessment Strategy

Assessment strategies may focus on the development of mission statements and benchmark tools alone, or they may include an analysis of the ways that institutional officials use such tools to influence faculty adherence to responsible research practices. Assessment efforts can also be used to review compliance strategies (including the compliance of faculty and research staff), student surveys, and sponsor evaluations, as well as to analyze rewards, incentives, and penalties.

In one application of a performance-based assessment, for example, Gilmer (1995) has the students in her research ethics course at Florida State University develop individual portfolios. The students must define their goals for learning and then present several pieces of their work that demonstrate that such learning has been accomplished. Portfolio assessment has been used primarily in educational settings to document the progress and achievements of individual students or teachers, but it has the potential to be a valuable tool for program assessment as well.

Another example is the Quality Improvement Program instituted by the Office of Human Research Protections at the Department of Health and Human Services (DHHS) (OHRP, 2002). This voluntary program is intended to help institutions prepare for and successfully achieve accreditation of their human research protection programs by private accrediting agencies. One feature of this program is a Quality Assurance Self-Assessment Tool, which helps the institution determine its level of compliance with federal regulations. In a related effort, the Health Improvement Institute (a private foundation), under a contract with DHHS, has established a national Award for Excellence in Human Research Protection, and the institute currently is developing performance-based evaluation criteria for the award (HII, 2002).

On a more general basis, a classic example of performance-based assessment is the Government Performance and Results Act of 1993 (Pub. L. No: 103–62), which requires federal agencies to develop strategic plans for delivering high-quality services or products to the public. Each agency has to (1) establish top-level agency goals and objectives, as well as annual program goals; (2) define how it intends to achieve those goals; and (3) demonstrate how it will measure agency and program performance in achieving those goals (NPR, 1997). As part of this initiative, the National Performance Review (NPR) was created to sponsor and organize benchmarking studies aimed at making government work better and cost less.

Performance-based assessment and benchmarking also are the basis for the Malcolm Baldrige Quality Award. Established by the Malcolm Baldrige National Quality Improvement Act of 1987 (Pub. L. No. 100–107), the award is administered by the National Institute of Standards and

Technology (NIST) and recognizes U.S. organizations for their achievements in quality and performance. The awards are given to businesses (manufacturing, service, small business) and to education and health care organizations (Seymour, 1995). Judging is based on seven criteria: leadership, strategic planning, customer and market focus, information and analysis, human resource focus, process management, and business results (NIST, 2001). The Baldrige criteria are used by thousands of diverse organizations for self-assessment and training and as a means to develop performance and business processes.

Several Institute of Medicine (IOM) reports describe performance-based approaches to improving community health (IOM, 1996a, b, 1997).

Strengths

Performance-based systems are increasingly common in diverse institutional settings, including health care (with the new emphasis on quality), the transportation sector, and various sectors of the manufacturing and service industries. Conceptual frameworks, measurement tools, and institutional case studies exist that can provide the foundation for the development of such a system in the area of integrity in research (see Appendix B). The setting of benchmarks represents a visible and tangible public commitment to integrity in research, and at the same time recognizes the need for oversight and assessment of questionable conduct in the research environment. Such goals can be flexible and consistent with the diverse institutional cultures of different research centers. Rewards, incentives, and penalties offer valuable educational resources in that they can be used to demonstrate the types of conduct that research institutions do and do not encourage.

Limitations

Performance-based systems also have limitations. They require a considerable amount of institutional commitment and involvement. Institutional officers need to exercise leadership and authority in the development of a mission statement and performance goals, as well as in the selection of benchmarks that will be used to guide and evaluate behavior. The adoption of performance-based goals can be divisive and controversial if faculty do not share common norms and aspirations, or if such goals lead to restrictions on the types of research that can be conducted. Institution-directed performance goals may run counter to the autonomy and innovative spirit of the research environment in most academic centers. Current limitations, which can be addressed through further research, include the need to identify the benchmarks and criteria most relevant to integrity in research, and the need to develop and validate the instruments and measures for assessment.

AN INSTITUTIONAL SELF-ASSESSMENT AND PEER REVIEW APPROACH

The complexities of balancing formal and informal approaches to fostering integrity in research have led to efforts to have research institutions assess their own performances, including the performances of their managements, faculty and research staff, in terms of complying with stated standards, goals, and practices. Such self-assessments may include evaluations of aspects of certification or institutional assurance of compliance with professional standards within a broader organizational context; this practice is frequently used in the accreditation of professional schools and departments, as well as of educational institutions. This approach is described in detail in Chapter 6. It is presented here only briefly to provide a basis for comparison with the other approaches discussed above.

Assessment Strategy

The strategy used in the self-assessment and assurance framework has multiple distinct features. The self-assessment process may be voluntary or mandatory. The process of institutional assurance of compliance with professional standards may be linked to an informal peer-review process or may be part of a formal accreditation procedure.

Different institutions will draw on one or more of these features, according to their own goals and resources. One such approach is presented for purposes of illustration.

The Middle States Commission on Higher Education (MSCHE, 2002) has recently developed proposals for new accreditation processes for its member institutions. The proposals include several distinct features of the internal assessment and accreditation process, including the following:

- institutional self-study;
- a team visit;
- types of accreditation actions;
- periodic review reports;
- institutional profile (annual) reports;
- candidacy and initial accreditation procedures;
- public information;
- use of technology (e.g., electronic submission of report materials); and
- training of evaluators and the institutions' departmental chairs.

The accreditation proposals are intended to address the commission's desire to balance the need to be careful and thorough with competing concerns about the slow and time-consuming nature of self-assessment and accreditation procedures (personal communication, J. Morse, Middle State Commission on Higher Education, [January 4, 2002]; see also www. msache.org/chx02.pdf). Additional examples of assessment strategies are described in Appendix B.

Strengths

The mission statements of accreditation bodies frequently assert that their role is to promote academic quality through formal recognition of the importance of compliance with professional standards and to advance the process of self-regulation (see, for example, the mission statement of the Council for Higher Education Accreditation [CHEA, 2001]).

Institutional self-assessment and accreditation procedures can be powerful influences in the shaping of professional behavior within a self-regulatory system. In the mid-1990s, for example, the National Council of Examinations for Engineers and Surveyors introduced engineering ethics questions in its examination used for granting professional engineering licenses. Every state and U.S. territory now uses these examinations. The inclusion of ethics questions in the engineering examination has been a practical force for introducing these issues into the classroom as well as into the review sessions that are popular with applicants for professional engineering licenses (Rabins, 1998).

If self-assessment and assurance requirements are implemented through standard practices used by faculty, then they can become important components in the overall climates of research organizations. In this way, they will have the ability to change institutional cultures over time. The process of self-regulation has the advantage of focusing on those areas that the members of the profession believe are essential to quality and integrity.

The introduction of elements of integrity in research and the responsible conduct of research into ongoing processes of institutional self-assessment and assurance helps accustom researchers to addressing these issues as part of the normal accreditation process. It thus diminishes the growth of disparate units within research organizations and strengthens the alignment of concerns about integrity in research with research practices.

Limitations

As noted above in the example of MSCHE practices, self-assessment and accreditation practices are frequently slow and time-consuming. Resources for the training of faculty evaluators and the operation of review committees are required at multiple levels within research institutions, including individual departments, research centers, and general administrations. Turnovers in personnel, a lack of familiarity with review and assessment procedures, and changes in the standards that are used as the basis for self-assessment and evaluation can further complicate the cumbersome nature of the assessment process.

At this time, an obvious limitation to the self-assessment approach as a means to foster integrity in research is the need to develop and validate evaluation instruments and measures. Further research on methods and measure and elements of the research environment (as described in Chapter 7) should eliminate this limitation. Examples of relevant assessment tools that might be adapted to the research environment are described in Appendix B.

SUMMARY

Evaluations of activities within research institutions occur in diverse forms and are influenced by different approaches that may consist of voluntary or mandatory elements and that may rely upon professional or volunteer reviewers. The committee has not found research evidence that suggests that any particular approach produces significant differences in measurable outcomes.

Each approach has certain strengths as well as limitations. The regulatory compliance approach may be more desirable when professional standards are clear and when measurable indicators can be developed to assess levels of compliance with selected processes and rules in different research settings. However, this approach often fosters attention to minimal standards (the lowest-common-denominator approach) rather than encouraging institutions to determine what is right for their situation and to invest in the efforts necessary to foster more desirable outcomes.

Although benchmarks are not yet available to support performance monitoring in the area of research integrity, they could be developed through educational programs and consensus building efforts. The adoption of a performance-based approach enables institutions to move beyond procedural compliance mechanisms as a self-regulatory device (e.g., certifying that they have adopted certain policies, procedures, and educational training efforts) in favor of a framework that fosters greater individuality but still adheres to certain performance standards that reflect basic peer and community values.

The institutional self-assessment approach is frequently overlooked in policy debates about research oversight. This approach is commonly associated with highly specialized efforts, such as the accreditation of academic institutions or procedures for the handling of laboratory animals. Recent concerns about the treatment of human subjects in research studies have brought new attention to the strategies involved in institutional self-assessment and assurance procedures (IOM, 2001; NBAC, 2001). The committee believes that much of this analysis is relevant to discussions of integrity in research, and therefore the committee suggests that more work is needed to determine how self-assessment coupled with peer review, particularly in the context of institutional accreditation, can be adapted to efforts to foster integrity in research. Chapter 6 provides a fuller discussion of this method of assessment.

REFERENCES

Burman WJ, Reves RR, Cohn DL, Schooley RT. 2001. Breaking the camel's back: Multicenter clinical trials and local institutional review boards. *Annals of Internal Medicine* 134:152–157.

CHEA (Council for Higher Education Accreditation). 2001. *Directory of CHEA Participating and Recognized Organizations, 2000–2001.* [Online]. Available: http://www.chea.org/Directories/index.cfm [Accessed February 25, 2002].

Gilmer PJ. 1995. Teaching science at the university level: What about the ethics? *Science and Engineering Ethics* 1:173–180.

HII (Health Improvement Institute). 2002. *Award for Excellence in Human Research Protections.* [Online]. Available: http://www.hii.org/ [Accessed May 14, 2002].

IOM (Institute of Medicine). 1996a. *Using Performance Monitoring to Improve Community Health: Conceptual Framework and Community Experience.* Washington, DC: National Academy Press.

IOM. 1996b. *Using Performance Monitoring to Improve Community Health: Exploring the Issues.* Washington, DC: National Academy Press.

IOM. 1997. *Improving Health in the Community: A Role for Performance Monitoring.* Washington, DC: National Academy Press.

IOM. 2001. *Preserving Public Trust.* Washington, DC: National Academy Press.

MSCHE (Middle States Commission on Higher Education). 2002. Characteristics of Excellence in Education: Eligibility Requirements and Standards for Accreditations. [Online]. Available www.msache.org/chx02.pdf [Accessed March 18, 2002].

NBAC (National Bioethics Advisory Commission). 2001. *Ethical and Policy Issues in Research Involving Human Participants.* Rockville, MD: U.S. Government Printing Office.

NIH (National Institutes of Health). 1999. *NIH Initiative to Reduce Regulatory Burden.* [Online]. Available: http://grants1.nih.gov/grants/policy/regulatoryburden/ [Accessed May 6, 2002].

NIST (National Institute of Standards and Technology). 2001. *Getting Started with the Baldrige National Quality Program: Criteria for Performance Excellence.* [Online]. Available: http://www.quality.nist.gov/ [Accessed May 15, 2002].

NPR (National Performance Review). 1997. *Serving the American Public: Best Practices in Performance Management.* [Online]. Available: http://govinfo.library.unt.edu/npr/library/papers/benchmrk/nprbook.html [Accessed May 15, 2002].

Nature. 2001. Time to cut regulations that protect only regulators. *Nature* 414:379.

OHRP (Office of Human Research Protections). 2002. *Quality Improvement Program.* [Online]. Available: http://ohrp.osophs.dhhs.gov/humansubjects/qip/qip.htm [Accessed May 14, 2002].

Rabins MJ. 1998. Teaching engineering ethics to undergraduates: Why? what? how? *Science and Engineering Ethics* 4:291–302

Seymour D. 1995. *The AQC Baldrige Report: Lessons Learned by Nine Colleges and Universities Undertaking Self-Study with the Malcolm Baldrige National Quality Award Criteria.* Washington, DC: American Association for Higher Education.

Sugar AM. 2002. Letters: The crisis in local institutional review boards. *Annals of Internal Medicine* 136:410–411.

Victor, B. 2001. *Integrity in the business environment.* Presentation at the June 28, 2001, meeting of the Institute of Medicine Committee on Assessing Integrity in Research Environments, Washington, DC.

5

Promoting Integrity in Research through Education

For many institutions, the impetus for the development of educational programs in the responsible conduct of research came from the 1989 mandate of the National Institutes of Health (NIH) to provide such education to all graduate students and postdoctoral fellows supported by National Research Service Awards (NIH, 1989). However, in this chapter, the committee argues that the provision of instruction in the responsible conduct of research derives from a premise fundamental to doing science: the responsible conduct of research is not distinct from research; on the contrary, competency in research entails responsible conduct and the capacity for ethical decision making. Indeed, the committee argues that integrity in research should be developed in the context of an overall research education program. The committee believes that doing so will be the best way to accomplish the following five objectives:

1. emphasize that responsible conduct is central to conducting good science;
2. maximize the likelihood that education in the responsible conduct of research influences individuals and institutions rather than merely satisfies an item on a "check-off" list for that institution;
3. impart essential standards and guidelines regarding responsible conduct in one's discipline;

4. enable participants in the educational programs to develop abilities[1] that will help them to effectively manage concerns related to responsible conduct of research as they arise in the future; and

5. verify that the first four objectives have been met.

The committee believes that useful insight into the best practice for education in the responsible conduct of research comes by analogy to the education of students in the critical analysis of the research literature in their fields. How is critical reading taught? First, students are introduced to the primary literature as soon as they enter an educational program. Second, the complexity of the readings and the depth of the analysis are gradually increased. Third, critical reading of journal articles, under the guidance of a mentor, is integrated into all aspects of the curriculum: core courses, seminars, the design of research projects, and the preparation of research manuscripts. Fourth, critical reading is taught by the very scientists who provide instruction in other aspects of research and who serve as primary role models. Finally, student competence is tested whenever students are asked to provide support for their ideas and conclusions. Consistent with the principles of effective instruction, assessment and feedback are continually provided from a student's first seminar presentation to the final thesis defense and submission of manuscripts for publication.

Education in the responsible conduct of research should be no less integral to the education of a researcher (Fischer and Zigmond, 1996; Gifford, 1994; Hensel, 1991). This principle was adopted by the National Academy of Sciences in 1992: "Scientists and research institutes should *integrate into their curricula* educational programs that foster faculty and student awareness of concerns related to the integrity of the research process" (emphasis added) (NAS, 1992). Moreover, when this committee advocates the promotion of integrity in the institutional research environment, it is advocating the creation of a climate in the institution, the department, and the research group that promotes integrity in research.

The committee recommends a model for education in the responsible conduct of research that includes the following principles:

---

[1]Abilities are complex combinations of motivations, dispositions, attitudes, values, knowledge of concepts and procedures, skills, strategies, and behaviors. These combinations are dynamic and interactive, and they can be acquired and developed through both education and experience (Mentkowski, 2000).

1. The educational program should be built around the development of abilities that give rise to responsible conduct. These include the ability to (a) identify the ethical dimensions of situations that arise in the research setting and the laws, regulations, and guidelines governing one's field that apply to those situations (*ethical sensitivity*); (b) develop defensible rationales for a choice of action (*ethical reasoning*); (c) integrate the values of one's professional discipline with one's own personal values (*identity formation*) and appropriately prioritize professional values over personal ones (*showing moral motivation and commitment*); and (d) perform with integrity the complex tasks (e.g., communicate ideas and results, obtain funding, teach, and supervise) that are essential to one's career (*survival skills[2]*).

2. The program should be designed in accordance with basic principles of adult learning. In particular, education in the responsible conduct of research should (a) be provided within the context of the overall education program, including adviser-trainee interactions, the core discipline-specific curriculum, and explicit education in professional skills; (b) take place over an extended period of time, preferably the entire educational program, and include review, practice, and assessment; and (c) involve active learning, including interactions among the instructors and the trainees.

3. The instruction should be provided as much as possible by faculty who are actively engaged in research related to that of the trainees.

This chapter is divided into four sections. It begins by briefly discussing abilities that should form the basis of education in the responsible conduct of research. It then outlines some of the emerging principles of adult learning. Next, it discusses how one might develop an effective curriculum, including how best to make use of the approaches now being used at many institutions. The final section summarizes the committee's findings.

Educational efforts on the responsible conduct of research should be designed to reach everyone involved in scientific research. As noted in Chapter 2, institutional leaders (e.g., chancellors, presidents, deans, chief executive officers) set the tone for the institutions with their own actions. Similarly, research leaders set an example with their own research practices. As discussed in Chapter 3, evidence from the organizational research literature demonstrates a relationship between supervisor behavior and the ethical conduct of the members of an organization (Posner and Schmidt, 1982, 1984; Walker et al., 1979). Continuing education of senior

---

[2]Here the term *skills* is not used in the narrow sense that suggests a dichotomy between knowing and doing.

researchers and administrators demonstrates a commitment of leadership to integrity in research and may help close the gap between what is taught and what trainees and junior staff see in practice (Hafferty and Franks, 1994; Hundert, 1996). Without formal training for all existing researchers and an instructional program for new staff and researchers, an institution will not be able to develop a consistent message to trainees and students.

CREATING A LEARNING ENVIRONMENT THAT FOSTERS INTEGRITY IN RESEARCH

To create a learning environment that fosters integrity in research, educators need to consider what is known about the development of integrity in other professional contexts and what that information suggests about the abilities that enable responsible conduct. A substantial body of literature drawn from a variety of research traditions (Rest, 1983) indicates that whether professionals engage in responsible professional conduct depends on the developmental abilities briefly described in point 1 above and explained more fully in the sections that follow.

Research also demonstrates that individuals participating in a formal educational program and seasoned professionals can be influenced by an educational environment that provides opportunities to develop the four abilities mentioned in point 1 above (Bebeau, 2001). (For operational definitions of each of the psychological processes from which the abilities are defined, see the Four-Component Model of Morality in Box 5-1.) The processes related to ethical decision making consider that each of the four components is a mix of affective and cognitive processes that contribute to the component's primary function (Bebeau et al., 1999; Rest, 1983) (see Chapter 3). The implication, then, is to teach the abilities (derived from these psychological processes) in context, as proposed in the sections that follow.

Interpreting the Ethical Dimensions of Problems in the Research Setting

Ethical sensitivity involves the awareness by researchers of how their actions affect others. In addition to the ability to anticipate the reactions and feelings of colleagues, supervisors, research participants, and others, ethical sensitivity involves being aware of alternative courses of action and how each could affect the parties concerned. It also involves the ability to construct possible scenarios with knowledge of cause-consequence chains of events in the research environment. Ethical sensitivity requires empathy and role-taking skills. For individuals being socialized to the research setting, ethical sensitivity involves the ability to see things

---

BOX 5-1
The Four-Component Model of Morality

Starting from the question "How does moral behavior come about?" Rest (1983) suggested that the literature supports at least four component processes, all of which must be activated for moral behavior to occur. These four components are:

Moral sensitivity. Moral sensitivity (interpreting the situation as moral) is the awareness of how one's actions affect other people. It involves being aware of the different possible lines of action and how each line of action could affect the parties concerned (including oneself). Moral sensitivity involves imaginatively constructing possible scenarios (often from limited cues and partial information), knowing cause-consequence chains of events in the real world, and having empathy and role-taking skills. Moral sensitivity is necessary to become aware that a moral issue is involved in a situation.

Moral judgment. Once a person is aware that various lines of action are possible, one must ask which line of action is more justified morally. This is the process emphasized in the work of Piaget (1932) and Kohlberg (1984). Even at an early stage in life, people have intuitions about what is fair and moral and make moral judgments about even the most complex of human activities.

Moral motivation and commitment. Moral motivation and commitment involves prioritization of moral values over other personal values. People have many values (e.g., values related to their careers, affectional relationships, aesthetic preferences, institutional loyalties, hedonistic pleasures, and things that excite them).

Moral motivation and moral character and competence. Moral character and competence is having the strength of your convictions, having courage, persisting, overcoming distractions and obstacles, having implementing skills, and having ego strength. A person may be sensitive to moral issues, have good judgment, and prioritize moral values; but if he or she is lacking in moral character and competence, he or she may wilt under pressure or fatigue, may not follow through, and may be distracted or discouraged, and moral behavior will fail. This component presupposes that one has set goals, has self-discipline and controls impulses, and has the strength and skill to act in accord with one's goals.

It is noteworthy that the model is not conceived as a linear problem-solving model. For example, moral motivation may affect moral sensitivity, and moral character may constrain moral motivation. In fact, Rest (1983) makes clear the interactive nature of the components. Furthermore, and in contrast to other models of moral function that focus on the traditional three domains—cognitions, affect, and behavior (Eisenberg, 1986; Lickona, 1991)—the Four-Component Model of Morality assumes that cognition and affect co-occur in all areas of moral functioning. Thus, moral action is not simply the result of separate affective and cognitive processes operating as part of an interaction. Instead, each of the four components is a mix of affective and cognitive processes that contribute to the component's primary function (e.g., identifying a situation as moral). Bebeau and colleagues (1999) suggest that researchers focus attention on identifying processes as they contribute to moral action rather than attempting to understand moral actions from a starting point defined by arbitrarily dividing moral functioning into cognitions, affect, and behavior.

SOURCE: Adapted from Bebeau et al. (1999).

from the perspective of other individuals and groups (including other cultural and socioeconomic groups), and, more abstractly, from legal, institutional, and national perspectives. Thus, it includes learning the laws, regulations, guidelines, and norms of one's profession and recognizing when they apply. In professional settings, the focus is on ethical sensitivity, rather than the more general "moral sensitivity" described in the operational definition (Box 5-1), to signal the distinctive expectations of the researcher that derive from the norms and rules that govern research practice.

Research on ethical sensitivity in professional settings indicates that (1) ethical sensitivity can be reliably assessed, (2) students and professionals vary in their sensitivities to ethical issues, (3) ethical sensitivity can be enhanced through instruction, and (4) the sensitivity to issues is distinct from the ability to reason about issues (Bebeau, 2001). See Appendix B for a more extensive discussion of the findings from several professions that have studied ethical sensitivity in relationship to professional performance.

Teaching Strategies

Many educators are familiar with sensitivity training that addresses such topics as affirmative action, gender equity, multiculturalism, awareness of diversity, and sexual harassment; and each of these topics has an appropriate place in the research setting. However, to promote training in ethical sensitivity in the responsible conduct of research, one also needs to focus on essential policies and practices related to the conduct of research. Such issues include the use of humans and animals in research; rules and codes governing environmental health and safety; processes and procedures for dealing with allegations of misconduct; authorship policies and practices; the acquisition, management, sharing, and ownership of data; conflicts of interest and commitment; and the responsible management of grant funds (see Chapter 2).

It is not that learners need to memorize policy documents and pass multiple-choice tests to demonstrate the acquisition of knowledge about details related to each of the content areas. Indeed, focusing on such details is often what learners view as demeaning. However, students need to know that such policies and guides exist and why they exist. In instructional settings they should be referred to often enough that students become familiar with them and references to them become habitual. To engage students in familiarizing themselves with the policies and practices, educators can rely on the techniques advocated in problem-based learning.

Designing Cases

Educators can design real or hypothetical situations that require learners to refer to policy guides as they identify stakeholders, consider consequences, and engage in probabilistic reasoning. What distinguishes sensitivity training from other kinds of case analysis is the way in which the instructor presents the material used to promote discussion. Distinct from the cases typically used in ethics courses, the information used in cases designed to foster ethical sensitivity is not predigested or interpreted. Instead, the case merely presents clues to a problem without signaling the particular violation of interpersonal, cultural, or normative practices that is being exhibited in the material. Through the use of such cases, learners can be directed to institutional policies and professional guidelines that set forth appropriate behavior. The challenge in a sensitivity assessment often is to distinguish the relevant information from the irrelevant information, to recognize the norms and values that should be considered, and even to recognize when these norms, rules, and values have been violated.

Assessment Methods

Tests of ethical sensitivity have been developed in a variety of professional settings (see Appendix B). These tests often involve the same types of cases that are used for instructional purposes and might require a student to witness on either videotape or audiotape an interaction that replicates professional interactions and that provides clues to a professional ethical dilemma (Bebeau and Rest, 1990). For example, the Racial Ethical Sensitivity Test (Brabeck, 1998) consists of five videotaped scenarios that portray acts of intolerance exhibited by professionals in school settings. Each scenario includes five to nine acts of intolerance that violate one or more of the common principles specified in codes of ethics.

Developing Competence in Reasoning About the Complex Problems That Arise in the Research Setting

Ethical reasoning implies that professionals be able to critically analyze their own moral arguments and develop defensible points of view for new problems that are likely to emerge during the course of professional life. Studies of the reasoning development of individuals in a variety of professions (Rest and Narváez, 1994), including students in research training (Heitman et al., 2000), indicate that persons entering a profession are not equally able to apply moral ideals to the resolution of complex moral issues. In fact, some novices and even some experienced professionals are unable to reliably distinguish simplistic moral arguments that appeal to self-interest and the maintenance of interpersonal affiliations as guides for moral action (referred to as a "personal interest moral schema") from arguments that appeal to societal laws and rules as a basis for deciding what should be done (a "maintaining norms moral schema"). In contrast, some professional students and research trainees are as able as persons with training in moral philosophy to work out what should be done in circumstances in which conflicts of interests exist. The mark of mature moral reasoning is the ability to figure out how to fairly modify existing rules or laws to accommodate the new moral problem that has emerged (referred to as "postconventional moral thinking") (Rest et al., 1999).

The effects of ethics instruction on a professional's moral reasoning has been extensively studied (Rest and Narváez, 1994). In medicine, for example, Self and Baldwin (1994) have reviewed a wide range of studies that have used the Defining Issues Test (DIT) or other measures of moral judgment to assess reasoning development. They concluded that (1) a medical curriculum without an ethics curriculum tends not to enhance moral reasoning; (2) instruction can be effective, although not all interventions produce significant gains; (3) the effects of at least some interventions can be attributed to an intervention based on comparisons with control groups; (4) strategies other than discussion of a dilemma can produce change; and (5) there is a relationship between reasoning and a range of indicators of physician performance.

Although other intervention strategies can produce a change in reasoning, the most consistent effects in professional education have been achieved with a teaching and assessment strategy that incorporates the dilemma discussion technique (for example, significant change for 14 of 15 cohorts of dental students [Bebeau, 2001]). Over a 10-year period (1983 to 1993), Bebeau and colleagues tested the strategy, incrementally adding instructional elements to improve students' ability to develop well-reasoned written arguments for addressing solutions to problems that students commonly encounter. The greatest improvements were achieved when students were provided with criteria for judging the adequacy of arguments and multiple opportunities to develop well-written arguments both before and after case discussions and when they received feedback on the strengths and shortcomings of their arguments from peers as well as from the instructor (Bebeau, 1994).

In a reexamination of intervention effects, Rest and colleagues (1997) found two different effects of an intervention. One was the acquisition of new thinking (increases in preferences for postconventional arguments—the effect that researchers have typically reported); the second effect was systematic rejection of simplistic thinking (decreases in preferences for personal-interest arguments). Rest and colleagues concluded: "From a practical educational point of view, both kinds of developmental progress are desirable: gaining more sophisticated moral thinking and also becoming clearer about what ideas should be totally rejected for their simplistic and biased solutions" (Rest et al., 1997, p. 500). As a consequence, researchers using DIT are encouraged to assess interventions in terms of moral judgment profiles (i.e., the proportion of arguments that appeal to each of the moral schemas) rather than just in terms of advances in post-conventional thinking (Bebeau, 2001).

In the early 1990s, researchers at the Poynter Center designed *Moral Reasoning in Scientific Research: Cases for Teaching and Assessment* (described below), a series of cases for teaching and assessment that incorporated the instructional techniques first tested in dentistry. When experienced researchers review these types of cases, as well as others included in available casebooks, they may judge the cases as too simplistic and be tempted to discard them in favor of discussions around contemporary issues that present highly challenging dilemmas. A danger in limiting teaching of responsible conduct to a discussion of contemporary cases is that students may learn the rules for specific situations but not be able to generalize to other issues of immediate relevance. Moreover, by focusing entirely on contemporary cases, students may not acquire the skills needed to identify the fallacies in their own arguments or to deal with many of the future unanticipated issues. Thus, the committee encourages faculty to develop a curriculum that provides opportunities to apply the more general moral reasoning ability that develops as a result of higher education to the specific problems that arise in the research setting (i.e., to develop "ethical reasoning"). On the basis of the original work of Kohlberg (1984) that was expanded by Rest, Bebeau, and colleagues (Bebeau, 1994; Bebeau et al., 1995; Rest et al., 1986), the committee defines ethical or moral reasoning as the ability to systematically examine a situation and then choose and defend a position on the issue (Bebeau et al., 1995). Arguments are evaluated in terms of the respondent's ability to describe the following:

- ethical issues and points of conflict, including precedents, principles, rules, or values that support prioritizing one interest over another;
- the stakeholders or parties that have a vested interest in the outcome of the situation;
- the probable consequences of possible courses of actions; and
- the ethical obligations of the central characters.

There is an important distinction between the focus in the development of cases designed to promote the sensitivity just discussed and those designed to promote ethical reasoning. Unlike cases for ethical sensitivity—in which finding and understanding the conflict (i.e., becoming sensitive to the conflict) is the focus—with ethical reasoning one is presenting a conflict that is relatively easily identified and interpreted. It has been shown that instruction in ethical reasoning is effective in increasing the ability of emerging professionals to engage in such tasks (Bebeau, 2001).

Teaching Strategies

To ensure that learners engage in reasoning about moral issues rather than in problem solving, a case description is followed by the question "Should the protagonist ___?" (e.g., take the data from the research setting or add an author to a manuscript) rather than "What should the protagonist do?" Learners are asked to take a tentative position either in favor or against the proposed action and to develop the best argument possible. To ensure that discussions are not just windy exchanges of opinion, the course facilitator is encouraged to have students explore the criteria for judging moral arguments before engaging in discussion and then use the criteria to critique each other's verbal or written arguments.

Assessment Methods

Tools that can be used to assess competence in ethical reasoning are available. Two that are well validated and suitable for adults have already been mentioned: *Moral Reasoning in Scientific Research: Cases for Teaching and Assessment* (Bebeau et al., 1995) and the Defining Issues Test (DIT) (Rest, 1979; Rest et al., 1999). In the former, the case studies are designed to facilitate improvements in reasoning as well as to assess such improvements. Each case is accompanied by extensive notes and checklists to help the evaluator achieve reliable judgments. The latter (Rest, 1979; Rest et al., 1999) is a paper-and-pencil measure of moral judgment based on Kohlberg's (1984) pioneering work on the development of lifespan moral judgment. DIT measures the reasoning strategies (moral schemas) that an individual uses when confronted with complex moral problems, as well as the consistency between reasoning and judgment.

More extensive descriptions of these tools, including data on their validity, are included in Appendix B. Many other collections of case studies also exist that could be used directly or modified slightly to serve as cases for teaching and assessment of ethical reasoning (see the section Responsible Scientific Conduct in Appendix D).

Moral Motivation and Identity Formation

The third component in the Four-Component Model of Morality (Box 5-1) acknowledges that individuals have a number of legitimate concerns that may not be compatible with the moral choice. Financial and career pressures, established relationships, and idiosyncratic personal concerns, among many others, compete for the researcher's attention. Blasi (1985) notes that people differ in how deeply moral notions penetrate their self-understanding and in the kinds of moral considerations that are judged to be constitutive of the self. In other words, *moral motivation* varies. This requires the attention of educators. Understanding that one is responsible provides the bridge between knowing the right thing to do and doing it.

Blasi (1985) and Kegan (1982) see *identity formation* as a lifelong developmental process. Recent work on identity formation based on Kegan's developmental theory (Forsythe et al., in press) suggests that at least 30 percent of graduates from West Point have not achieved key transitions in identity formation that would enable them to have the broad, internalized understanding of and commitment to codes in the responsible conduct of research and other professional standards. Whereas such individuals may see codes and professional standards as guides for behavior, they are likely to conform to the guides simply to garner rewards and avoid negative consequences, without achieving an understanding of their personal responsibility. Forsythe and colleagues concluded that "[professional] development programs will not be successful in instilling desired values in less mature [preprofessionals] unless the broad educational environment in which they operate promotes identity development toward a shared perspective on professionalism" (Forsythe et al., in press). Evidence from studies of role concept development in dentistry (Bebeau, 1994) support these observations.

Recent work on integrity in research is directing attention to the need for more formal efforts to socialize trainees and beginning researchers to professional expectations and values. For example, Braxton and Baird highlight the need to socialize researchers to the role of self-regulation, arguing that "doctoral study can be configured so that future scientists are prepared to participate in the deterrence, detection, and sanctioning of scientific wrongdoing" (Braxton and Baird, 2001, p. 593). The responsibility for self-regulation would be addressed as part of identity formation, whereas the actual skills would be taught as part of survival skill education (see below). The need for such socialization is further confirmed by Anderson's (2001) study of doctoral students' conceptions of science and its norms. She concludes: "The theme of individual, independent work that runs through these interviews suggests that students might not be subject to as much osmotic group socialization as many faculty assume. It is also clear that the channels by which socialization to the normative aspects of academic life are communicated are primarily informal. Calls for more formal, more deliberate approaches to normative socialization find support in the vagueness with which students conceptualize the norms that underlie academic research" (Anderson, 2001, p. 6).

Anderson's study is continuing and is expected to demonstrate the extent to which these norms develop in a academic setting.

Teaching Strategies

One of the chief objectives of *On Being a Scientist* (NAS, 1989, 1995a) and *Honor in Science* (Sigma Xi, 1986) is to convey the central values of the scientific enterprise. In an earlier era, such values were typically conveyed informally, through mentors and research advisers. Today, students can be reached in more formal ways. They can be exposed to lectures on the norms and values of science, such as Merton's norms, buttressed by discussions of the philosophy of science and such books as Grinnell's *The Scientific Attitude* (1992). They also can be encouraged to read stories about exemplary scientists to gain a sense of how such individuals have conceptualized their role and responsibilities as a scientist, a mentor, and a member of the larger society. Such a story, together with commentary, has been developed for an outstanding scientist in dentistry (Rule and Bebeau, 2001). Other examples, such as that portrayed in Djerassi's novel *Canter's Dilemma* (1989), can be effective tools for helping students develop their personal identities as scientists. In addition to lectures and discussions about the general norms of science, educators in particular disciplines will want to introduce learners to the code of professional conduct for the discipline.

Assessment Methods

Two assessment methods can be used to evaluate role concept development. One is to ask students at various stages of their education to write a short essay entitled "What does it mean to become a scientist?" Such an essay can be critiqued on the basis of the extent to which the norms and values that undergird the scientific enterprise are described. Each educator can develop his or her own criteria for assessment of the essay, based on the instruction provided on the topic and the discipline's ethical code (see Bebeau [1994] for an example). A second method, developed in other professional settings, is the use of a norm-referenced measure of role concept; that is, the extent to which the individual incorporates the norms and values of the profession into his or her identity.

Appendix B includes descriptions of such measures and their value in assessing outcomes of the effort to promote role concept development.

Developing Self-Regulation, Self-Efficacy, and Implementation Abilities Necessary for Effective and Responsible Research Practice

Fundamental to responsible conduct in any profession is the ability to perform the complex tasks of the discipline with integrity, i.e., to have acquired *survival skills*. When the committee defined integrity in research (Chapter 2), it defined an aspect of moral character and experience. The fourth component in the Four-Component Model of Morality (Box 5-1) attends to the importance of character to the effective and responsible conduct of research. Integrity, ego strength, perseverance, backbone, toughness, strength of conviction, and courage are also qualities required for effectiveness as a researcher. A researcher may be ethically sensitive, may make good ethical judgments, and may place a high priority on professional values; but if he or she wilts under pressure, is easily distracted or discouraged, or is weak willed, a moral failure may occur because of a deficiency in character and competence.

Professional educational programs assist individuals in understanding the broad fundamentals of their disciplines; gaining some depth in the details of a particular subarea; and obtaining practical experience in research, including experimental design, methodology, data analysis, and other practices detailed in the definition of integrity in research. Fischer and Zigmond (1998) point out that although graduate education has as its purpose the development of a range of practices relevant to the specialty, such programs often lack an essential dimension (Bloom, 1992; Widnall, 1991), the development of a set of general professional skills (Fischer and Zigmond, 1998; Magner, 2000; NAS, 1995b). For example, scientists should be able to present their results at scientific meetings, defend their chosen methodologies and interpretations of data, and prepare written reports. They need to be able to learn from critical comments and suggestions from their professional peers both at oral presentations and through peer review of manuscripts. They may need to obtain grants to fund their research, hire and supervise technical staff, and teach classes and advise individual students. Moreover, with the decrease in tenured positions in academia, science apprentices should learn how to find information on and prepare themselves for other types of careers (NAS, 1995b; Varmus, 1995). Morever, within this broader range of careers, skills other than those specific to the collection and analysis of data may become all the more essential (Cordes, 1997; Greenwood and Kovacs-North, 1999; NAS, 1995b).

Teaching Strategies

Individuals traditionally learn such professional skills in one of three alternate ways: through trial and error, from the teachings of their advisers, or through courses taught by faculty in the discipline from which the skill derives. However, each of these methods has its limitations. Trial and error can lead to professional fatalities, particularly among individuals without a great deal of experience with the culture of science. Education by individual advisers—important as it is (see below)—may be limited by the adviser's own education and resources. Courses devoted to a given skill (e.g., writing or oral presentations) may not be sufficiently relevant to the needs of a researcher or may require more time than is available in an already crowded curriculum. (For more discussion on these points, see Fischer and Zigmond [2001b].)

If one cannot count on these routes for gaining necessary skills, what other options are available? The University of Pittsburgh is one institution that has developed an educational model that provides science apprentices with an introduction to specific "survival skills." In that model, series of daylong workshops are run throughout the year, one per month for eight months, with each workshop being devoted to a specific topic (Fischer and Zigmond, 2001b). Many other models could be used, including an intensive minicourse on professional skills and a more traditional course meeting one to three times per week over one or more semesters.

Assessment Methods

Some aspects of survival skills might be assessed by simple tests; for example, trainees might be asked to edit a description of an experiment, evaluate a research article for possible publication, comment on a résumé, or communicate written feedback about some offense to a colleague. In each of these tests, the task should include something that requires attention to the ethical dimensions (e.g., a possible misrepresentation). In addition, whenever possible instructors should include a realistic, performance-based assessment. Trainees might be asked to assemble a portfolio that includes work that applies specific skills. Such work might include a manuscript, a videotape of a research seminar or class lecture, a poster, a grant application, or a plan for career development. Peers, faculty, or representatives of nonacademic careers could evaluate the portfolios (see Gilmer [1995, 2002] for examples of the use of portfolios). Faculty could use the definition of integrity in research (Chapter 2) to construct the criteria used to judge the contributions in the portfolio. Such an exercise would have the added benefit of subjecting trainees to the type of evaluation that many will experience as they move up the career ladder.

HOW DO ADULTS LEARN?

Having outlined four abilities that should be addressed in a program on the responsible conduct of research, it is also important to examine whether the methods proposed for teaching and assessing these abilities reflect an understanding of how people learn.

Principles of Adult Learning

This section briefly describes six learning principles that should be considered when developing an educational program on the responsible conduct of research.

1. *Education is best provided by individuals who have a deep understanding of their subject matter and whose teaching reflects that they care about and value the material being taught* (Wlodkowski, 1999). Teaching of the responsible conduct in research presents a special challenge because it requires a synthesis of ethics and science. When scientists and ethicists collaborate in the design and implementation of learning experiences, students come to appreciate the complexity of problems that arise in the practice of science. Furthermore, when instruction requires the application of norms (and the ethical theories that support them), values, and rules and regulations to the practical problems that arise in the day-to-day practice of science, learning is more likely to last and to transfer to new situations. It follows, then, that instruction in the responsible conduct of research by a team of faculty—or by a faculty member with expertise in both ethics and science—is optimal.

When faculty take time from their scholarly work to provide practical instruction that draws on expertise from related fields, they demonstrate the importance of this educational task and its relevance to the practice of science. Only faculty with a deep understanding of the complexities of the related disciplines can answer questions with concrete examples, avoiding the mistake that many so-called experts make in instructing adults: that simply knowing something is enough to teach it effectively (Wlodkowski, 1999). Turning over a course in the responsible conduct of research either to an ethicist with no understanding of the current practice of science or to a scientist who has not taken the time to educate himself or herself about the various processes involved in ethical decision making conveys to students that the subject matter under discussion is peripheral to the current realities in the field.

2. *Educational programs in responsible conduct of research should occur over an extended period; indeed, they should occur throughout a trainee's tenure*

at an institution. It is clear from educational research that "spaced review and practice" (learning that occurs over an extended period, with frequent opportunities for practice and feedback) leads to greater retention of information and longer-term behavior change than "massed learning" (learning that occurs intensively during a brief period). Furthermore, some learners take a great deal more time to learn something than others do. Academic learning time—that is, time spent actively and successfully involved in learning—is strongly related to achievement (Fisher et al., 1980). Such findings suggest that programs limited to several hours of instruction during a single day, or even spread over a few weeks, are unlikely to have much of a long-term impact.

3. *Active participation in problem-oriented learning is an important component of effective educational programs.* This principle argues for learning that is experiential and contextual. The teaching of rules and regulations and the testing of knowledge outside of the context in which they are applied not only violates this principle but also is devalued by self-directed learners who are well aware of the limits of learning that is disconnected from its application. Experiential learning enables participants to immediately apply and test their new skills, resulting in a more rewarding and effective learning experience (Darkenwald and Merriam, 1982; Dickenson and Clarke, 1975; Knowles, 1970; Tight, 1996).

Researchers have also found that adults are self-directed in their educational objectives (Cross, 1979; Even, 1981) and seek learning experiences that are directly applicable to their lives (Burgess, 1971; Carp et al., 1974). Moreover, giving students an opportunity to use language, orally and in writing, facilitates learning, both because of the active involvement inherent in such assignments and because it helps learners link prior knowledge to the new information that they are learning (Lemke, 1995). This principle argues for teaching to take place in small, interactive groups, and it further suggests the importance of problem-oriented assignments that require application of learning.

4. *Programs will be more effective if educators help students assess their prior knowledge and integrate new material with familiar ideas.* People bring different knowledge to the learning situation (von Glasersfeld, 1989), and individual learning involves transfer of that prior knowledge to the new situation (Tobin and Tippins, 1993). To encourage new learning, a teacher needs to help students assess their prior knowledge and integrate new material with familiar material. Teachers who learn what their students know, and who know how their students learn, are in a position to teach for conceptual change (Loucks-Horsley et al., 1998; Strike and Posner, 1992). Effective learning requires that trainees be knowledgeable about

themselves, and that they have frequent opportunities for assessment and feedback from instructors who are knowledgeable about their trainees.. This typically requires the use of small groups or, even better, individual conferences. In the context of the responsible conduct of research, it requires that trainees have opportunities to assess their strengths and shortcomings with respect to the broad abilities that contribute to competence in ethical decision making.

   5. *Students should be encouraged to share their own experiences with others in the class.* This not only seems to improve participant satisfaction but also provides a richer learning experience for all participants. Each student comes to the learning experience with factual knowledge as well as experiential learning connected with social roles, such as those associated with sex, race, class, and other affiliations. Course participants should be seen as de facto instructors as well as students. In this way, additional issues, information, and perspectives will be incorporated into the curriculum (Bruffee, 1993; Even, 1981; Merriam and Caffarella, 1999). This principle demands that some portion of a program of education in the responsible conduct of research involve interactions with and among the participants in a relatively small group.

   6. *Instructional programs that attend to developmental differences and individual learning preferences are more likely to be effective.* Individuals entering graduate education differ in their levels of intellectual development (King and Kitchener, 1994), moral development (Kohlberg, 1984; Rest, 1983), and identity formation (Blasi, 1985; Kegan, 1982). Developmental psychologists have shown how to construct learning experiences that promote developmental progress, and they have shown the negative consequences of learning and learner perceptions when instructional strategies are not appropriate to the learners' developmental levels.
   Although developmental differences are of prime importance for promoting the intellectual and moral dimensions of integrity in research, other individual differences also warrant consideration when designing educational programs. Furthermore, individuals vary in their preferred learning styles (Burgess, 1971; Dickenson and Clark, 1975; Knox, 1978). Many educators believe that learning is likely to be enhanced by accommodating different learning styles and different levels of intelligence through the use of a variety of teaching formats, including lectures and discussions, individual as well as group exercises, and both oral and visual stimuli (Parker and Rennie, 1998). Thus, a "one-size-fits-all" approach to education in the responsible conduct of research is not likely to be adequate.

Need for an Integrated Approach

Learning relies on the interconnection of four learning environments: the learner-centered, knowledge-centered, assessment-centered, and community-centered environments (Bransford et al., 1999; The Cognition and Technology Group at Vanderbilt, 2000). Learner-centered environments build on what the students bring to the classroom, their strengths, and their prior learning. Knowledge-centered environments help students construct new knowledge by providing organized disciplinary knowledge and the skills needed to use that knowledge. In assessment-centered environments, both students and teachers set goals, ask for feedback, and make revisions as needed. Community-centered environments establish the normative behaviors for individuals in the learning institutions and other aspects of their professional lives, as well as in the other communities of which they are a part.

Each of these learning environments influences the others (The Cognition and Technology Group at Vanderbilt, 2000; Mentkowski, 2000). Thus, an effective program of education in the responsible conduct of research requires a broad, multicomponent approach. The following section describes the key elements of this approach.

Incorporating Current Practices in an Effective Curriculum

The background section of this chapter outlined the committee's rationale for recommending that education in the responsible conduct of research be taught in the context of the overall educational program.

Formal education in the responsible conduct of research can be provided in a wide variety of ways. Some of the most common approaches are discussed here.

Adviser-Trainee Interactions

The research adviser typically plays a central role in discussions of many aspects of responsible conduct of research. Indeed, until recently, this was the major, if not the only, mechanism through which most trainees received such education. These discussions often include one-on-one meetings, research group meetings, and journal clubs that are led by the adviser. Interactions between a trainee and his or her adviser typically occur over a long period and can be individualized to the type of research being done, the regulations and guidelines that pertain to that research, and the specific needs of the trainee. Moreover, when the individuals who are principally responsible for instruction in research also play a key role in teaching about the responsible conduct of research, they indicate—by deed as well as by word, it is hoped—that they consider scientific integrity to be essential to doing good science.

Adviser-based instruction in the responsible conduct of research might involve two or more hours per month and thus could occupy well over 100 hours while a trainee is part of a research group. Thus, this single component could (and probably should) constitute the largest amount of time devoted to education in the responsible conduct of research for a given trainee.

As central and extensive as adviser-based education is, however, the committee does not consider this approach to be sufficient. First, if the instruction is limited to one-on-one instruction, the value of peer interactions may not be present. Second, some critical issues may not come up within a research group setting because they apply to considerations outside of the area of research, such as classroom teaching. Third, the research adviser may have little experience in mentoring or limited understanding of ways to promote integrity in research.

Short Courses

The short seminar seems to be the most common approach to formal education in the responsible conduct of research. Typically, a faculty member in philosophy or science will organize four to six sessions lasting 90 to 120 minutes each. The syllabus deals with what are deemed to be the essential issues for a given group of students (e.g., plagiarism, authorship, and ownership of data). The sessions often include some didactic material, such as an explanation of current conventions and a discussion of a case. The cases may be presented as written vignettes, films, or plays that are acted out by the participants. The same individual may give all lectures, or guest speakers may present the lectures. If the group is large (more than 15 to 25 participants), the discussions usually occur in smaller breakout groups. There may or may not be outside reading, and if grades are given, they are most often given on the basis of a paper that the participants write.

Such programs can provide a valuable component of education in the responsible conduct of research, provided the developers design them for purposes that relate to a comprehensive program. For example, a short course could be used to introduce new graduate students to the norms and values of the scientific enterprise. Alternatively, a course could be devoted to a particular topic, such as preparing an informed-consent protocol for a study of human subjects to be submitted to an institutional review board. Although short course are, by necessity, limited in scope, they may be able to address key issues in the responsible conduct of research in a multidisciplinary setting.

Full-Semester Courses

The model for a full-semester course is fundamentally the same as that outlined above for a short seminar. There may be an introductory unit on ethical theory, and the number of topics is significantly expanded. There also may be more reading and written assignments, as well as some quizzes or examinations. Typically, a variety of lecturers and discussion leaders teach the course. Such a course will involve one to three hours of instruction each week over 12 to 15 weeks—for a total of 12 to 45 hours.

The full-semester course is a step in the right direction, although many members of the committee believe that even this is too limited a time for the task of providing education in this important area. Some committee members prefer short seminars, given over multiple years, that can be tailored to the curriculum and the maturation of the student.

Single Workshops

Some programs elect to have a single workshop that lasts several hours and that focuses on a set of related topics (e.g., publication of research results). Students may be asked to attend several such sessions during the course of their education. Sometimes, given the extended length of a given session, lectures and breakout groups are combined with panels to provide a greater diversity of input. In contrast to the short seminars and full-semester courses, which may each involve 5 to 30 participants, workshops are often open to a relatively large number of students in a given educational program and sometimes draw several hundred participants. As with short seminars, however, the typically large numbers of individuals who participate in workshops often limit the amount of active learning or instruction that takes into consideration the knowledge base of individual students.

Computer-Based Instruction

As institutions work to provide education in the responsible conduct of research to ever-larger numbers of individuals, there seems to be an emerging use of computer-based educational programs, either via a centralized website or via diskettes or compact discs that are made available to individuals. These programs seem to focus most, if not all, of their attention on regulations and guidelines, and they often include an assessment of knowledge of that material. The programs may involve several different units, each of which takes one to three hours to complete, often at a single sitting.

Such programs often provide the least costly way to provide instruction to a large number of individuals. They also may be an effective way

to deal with certain kinds of education; for example, for providing familiarity with professional regulations and norms. On the other hand, they
typically are brief, involve little or no individualized instruction, and do
not provide researchers-instructors as visible role models. Furthermore,
although there may be some semblance of "active learning," it is typically
of a very primitive form.

Integration of Relevant Ethical Issues within the Core Curriculum

The committee believes that education in the responsible conduct of
research should be provided within the core curriculum of a discipline,
with ethics cases strategically selected to promote development of each of
the abilities that will enable responsible conduct. There are two reasons
for this. First, if responsible conduct is an integral part of conducting
research, as argued above, then it should be infused through the educational program for new researchers. Second, many topics do not fit logi-
cally into a more general context of the responsible conduct of research
but nevertheless deserve attention. Depending on the discipline, such
topics might include informed consent (for the use of human subjects in
research), the use of animals in research, data management, storage and
retrieval of data, and ethical issues related to developing technologies
(e.g., human cloning, gene therapy, and reproductive technologies).

Issues related to the responsible conduct of research can be integrated
into a core course in two complementary ways. First, individuals teaching
the courses can include comments on the ethical dimensions of a subject
as part of the lectures. Thus, a faculty member discussing genetic markers
for disease might comment on the ethics of genetic testing, whereas a
faculty member teaching a course in anthropology might mention the
possible adverse impact of fieldwork on the lives of indigenous people.
This can occur without any special fanfare but as a natural component of
the discussion of a given topic.

Second, time can be set aside to discuss an ethics case of particular
relevance to the focus of the course. This discussion might be included as
part of the series of breakout groups that are often organized for core
courses to focus on a particular research article. Faculty teaching in one or
the other formats can provide opportunities to learn to lead a discussion
that promotes ethical reasoning or to lead a group activity that promotes
ethical implementation (several texts with case studies are listed in Appendix D in the section Responsible Scientific Conduct).

Although the committee wholeheartedly supports the concept of integrating education in the responsible conduct of research into the core
curriculum, the committee also recognizes that accomplishing this is no
simple matter. One or two faculty members can mount their own educa-

tional program in the responsible conduct of research. But influencing the courses that other faculty members teach demands practical and time-consuming diplomacy. Moreover, it is an initiative that requires continual maintenance. There is often constant pressure on a course director to find time for new topics. In addition, the faculty and directors for a given course may change frequently. In both instances, issues of the responsible conduct of research can quickly disappear from the curriculum.

Florida State University, for example, offers an interdisciplinary science course (for graduate or undergraduate credit) that integrates the responsible conduct of research in science into each of its sections (Gilmer, 1995, 1998; Gilmer and Rashotte, Dec. 1989/Jan. 1990). The course uses historical examples, including the development of the atomic bomb, the Tuskegee syphilis trials, and scientific freedom and responsibility, to highlight the importance of science and the profound influence that science has on society. Current examples of integrity in science are seen within such a historical context. Assessment is a critical aspect of such a course, and writing is highlighted, with students reviewing and critiquing each other's ideas on a course website. Students use electronic portfolios to document their learning in the course (Gilmer, 1995, 2002). Students are given the option of selecting for one of their collaborative group projects a service activity that fits into the goals of the course. This way of bringing the consideration of integrity in science into the curriculum incorporates the four learning environments: the knowledge-centered, student-centered, assessment-centered, and community-centered environments (Bransford et al., 1999; The Cognition and Technology Group at Vanderbilt, 2000).

Ethics in the Context of Education in Professional Survival Skills

Earlier in this chapter, instruction in survival skills was described as one of the keys to the development of an environment that promotes integrity in research. The committee made this choice for three major reasons. First, virtually every aspect of any curriculum has an ethical dimension, and, as already noted, the committee believes that these issues are best identified and addressed in context. Professional skills, like more traditional aspects of a curriculum, have ethical dimensions. For example, when one is teaching about writing research articles, discussions might include issues of plagiarism, honorary authorship, data selection, and graphic design, while a workshop on grantsmanship might include a discussion of the importance of not overstating the sensitivities of methods or the quality of pilot data, exaggerating the assistance that one obtains from colleagues, or promising more than can be accomplished. As in the case of education in the responsible conduct of research within the core curriculum, the teaching of responsible conduct within a program of professional development can and should be done in two ways: through the inclusion of material in the lectures and through the active discussion of ethics cases.

Second, many of the ethical issues discussed in survival skills workshops are simply not likely to arise as topics either in the core curriculum or in traditional courses on the responsible conduct of research. Examples of such issues include the responsibility to publish worthwhile data in reports so that others may benefit from the work (particularly if the work was supported with public funds); the importance of acknowledging the contributions of others in oral presentations as well as in written work; and the responsibility to ensure that other researchers can replicate published results, by providing a complete and accurate presentation of methods and by being willing to share all reagents not commercially available.

The third reason stems from the fact that institutional climate appears to be an important determinant of responsible conduct in science (see Chapter 3). This may reflect, in part, the fact that most people are likely to learn less from what a faculty member or an institution offers as formal instruction than from the actual behavior that is observed. Offering instruction in survival skills is a clear indication that the faculty and the institution accept some responsibility for the professional advancement of their students. However, programs in professional skills should not replace or reduce the efforts of individual advisers to provide individual mentoring.

Other Venues

This brief discussion has not exhausted the ways in which instruction on the responsible conduct of research can be made an integral part of conducting scientific research. For example, authors should include issues related to integrity in science (including cases) in the textbooks of scientific disciplines (e.g., see Kovac, 1995; Tobin and Dusheck, 2001; and Zigmond et al., 1999) education directors should make ethical issues a component of annual retreats. Responsible conduct of research also should be a subject for online forums in areas of research (e.g., Fischer and Zigmond, 2001a; Zigmond and Fischer, 1995), and discussions of issues related to the responsible conduct of research should appear as part of the programs of professional meetings (for example, they are standard features of the meetings of the Society for Neuroscience and the American Association for the Advancement of Science). The objective is straightforward: to ensure that teaching of the responsible conduct of research exists side by side with discussions of all other aspects of science.

SUMMARY

In preparing science apprentices for success, curricula should address the broad range of skills that they will need as professionals. For example, they should be taught how to communicate their research data as well as how to collect them, how to teach as well as how to learn, and how to develop a career as well as how to develop a thesis. Moreover, the responsible conduct of research should be viewed as an integral part of good science and thus an integral part of education programs. It follows, therefore, that the objectives and the methods for the teaching of the responsible conduct of research should be nothing less than those used for the teaching of other skills and abilities valued within a discipline. Thus, the model for providing instruction in the responsible conduct of research is taken from traditional programs for teaching students what is necessary for their performance as researchers: (1) start as soon as the students arrive; (2) make the instruction in the responsible conduct of research part of everything they do, placing the education in the context of the research instead of making it a separate entity; (3) move from the simple to the complex; and (4) assess student competency. In this way, there is no mistaking the message: communicating well, obtaining employment and research grants, excelling in teaching, advising, and mentoring, engaging in ethical decision making, and behaving responsibly are at the core of being a researcher.

The committee finds that programs of education in the responsible conduct of research should aim to have an effect and should not be in place merely to be able to check an item off a list. They also should be based on current understanding of the psychological processes that give rise to morality and on current understanding of how adults learn.

Research advisers play a central role in the education of their trainees in the responsible conduct of research, not only by what they teach, but also by their own conduct. To facilitate this process, programs of adviser education and evaluation in this area are needed.

Adviser-based education of trainees should be supplemented by a program of education in the responsible conduct of research that is integrated into the overall educational program to include (1) a core course, (2) other specific educational program elements (e.g., journal clubs and retreats), and (3) individual research group meetings (e.g., laboratory meetings). Education in the responsible conduct of research should be built around the development of abilities that give rise to responsible conduct. Finally, education in the responsible conduct of research should involve research practitioners and individuals with expertise in ethics.

Although the field of assessment of the responsible conduct of research is still in a developmental stage, efforts to promote integrity in research need to be evaluated. Such an evaluation not only will provide useful information in the determination of the level of competence in that area but also will signal that integrity in research is a valued aspect of the educational program.

REFERENCES

Anderson M. 2001. *What Would Get You in Trouble: Doctoral Students' Conceptions of Science and Its Norms.* Proceedings of the ORI Conference on Research on Research Integrity. [Online]. Available: http://www-personal.umich.edu/~nsteneck/rcri/index.html [Accessed March 18, 2002].
Bebeau MJ. 1994. Influencing the moral dimensions of dental practice. In: *Moral Development in the Professions: Psychology and Applied Ethics.* Hillsdale, NJ: L. Erlbaum Associates Pp. 121–146.
Bebeau MJ. 2001. *Influencing the Moral Dimensions of Professional Practice: Implications for Teaching and Assessing for Research Integrity.* Proceedings of the ORI Conference on Research on Research Integrity. [Online] Available: http://www-personal.umich.edu/~nsteneck/rcri/index.html [Accessed March 18, 2002].
Bebeau MJ, Rest JR. 1990. *The Dental Ethical Sensitivity Test.* Minneapolis, MN: Division of Health Ecology, School of Dentistry, University of Minnesota.
Bebeau MJ, Pimple KD, Muskavitch KMT, Borden SL, Smith DL. 1995. *Moral Reasoning in Scientific Research: Cases for Teaching and Assessment.* Bloomington: Indiana University.
Bebeau MJ, Rest JR, Narvez DF. 1999. Beyond the promise: A perspective for research in moral education. *Educational Researcher* 28(4):18–26.
Blasi A. 1985. The moral personality: Reflections for social science and education. In: Berkowitz MW, Oser F, eds. *Moral Education: Theory and Application.* Hillsdale, NJ: L. Erlbaum Associates.
Bloom FE. 1992. Training neuroscientists for the 21st century. *Trends in Neurosciences* 15:383–386.
Brabeck MM. 1998. Racial ethical sensitivity test: REST videotapes. Chestnut Hill, MA: Lynch School Boston College.
Bransford JD, Brown AL, Cocking RR. 1999. *How People Learn: Brain, Mind, Experience, and School.* Committee on Developments in Science of Learning and Commission on Behavioral and Social Sciences and Education. Washington, DC: National Academy Press.
Braxton J, Baird L. 2001. Preparation for professional self regulation. *Science and Engineering Ethics* 7:593–614.
Bruffee KA. 1993. *Collaborative Learning: Higher Education, Interdependence, and the Authority of Knowledge.* Baltimore, MD: The Johns Hopkins University Press.
Burgess P. 1971. Reasons for adult participation in group educational activities. *Adult Education* 22:3–29.
Carp A, Peterson R, Roelfs P. 1974. Adult learning interest and experiences. In: Cross PK, Valley JR, eds. *Planning Non-Traditional Programs.* San Francisco, CA: Jossey-Bass.
The Cognition and Technology Group at Vanderbilt. 2000. Adventures in anchored instruction: Lessons from beyond the ivory tower. In: Glaser R, ed. *Advances in Instructional Psychology 5 (Educational Design and Cognitive Science).* Hillsdale, NJ: L. Erlbaum Associates.
Cordes C. 1997, October 24. Science Board considers federal role in improving graduate education. *Chronicle of Higher Education.* P. A32.
Cross P. 1979. Adult learners: Characteristics, needs, and interests. In: Peterson RE, ed. *Lifelong Learning in America.* San Francisco, CA: Jossey-Bass.

Darkenwald G, Merriam S. 1982. *Adult Education: Foundations of Practice.* New York, NY: Harper & Row.

Dickenson G, Clark KM. 1975. Learning orientations and participation in self-education and continuing education. *Adult Education* 26:3–15.

Djerassi C. 1989. *Canter's Dilemma.* New York, NY: Penguin Books.

Even MJ. 1981. The adult learning process. *Perspectives in Adult Learning and Development* 1:13–19.

Fischer BA, Zigmond MJ. 1996. Teaching ethics: Resources for researchers. *Trends in Neurosciences* 19:523–524.

Fischer BA, Zigmond MJ. 1998. Survival skills for graduate school and beyond. In: Anderson M, ed. *The Experience of Being in Graduate School: An Exploration (New Directions in Education Series, Number 101).* San Francisco, CA: Jossey-Bass. Pp. 29–40.

Fischer BA, Zigmond MJ. 2001a. *Map and Compass (an Electronic Column Sponsored by the International Brain Research Organization).* [Online]. Available: www.ibro.org [Accessed March 18, 2002].

Fischer BA, Zigmond MJ. 2001b. Promoting responsible conduct in research through "survival skills" workshops: Some mentoring is best done in a crowd. *Science and Engineering Ethics* 7:563–587.

Fisher C, Berliner D, Filby N, Marliave R, Cahen L, Dishaw, M. 1980. Teaching behaviors, academic learning time, and student achievement: An overview. In: Denham C, Lieberman A, eds. *Time to Learn.* Washington, DC: National Institute of Education.

Forsythe GB, Snook S, Lewis P., Bartone P. In press. Making sense of officership: Developing a professional identity for 21st century army officers. In: Snider D, Watkins G, eds. *The Future of the Army Profession.* New York, NY: McGraw-Hill.

Gifford F. 1994. Teaching scientific integrity. *The Centennial Review* 38:297–314.

Gilmer PJ. 1995. Teaching science at the university level: What about the ethics? *Science and Engineering Ethics* 1:173–180.

Gilmer PJ. 1998. Sowing the seeds of responsible conduct on new ground: The academic perspective. In: *The Responsible Conduct of Research: A Commitment for All Scientists,* Proceedings of a conference sponsored by the Public Responsibility in Medicine and Research, the Association of American Medical Colleges, and the National Institutes of Health. Boston, MA: PRIM&R. Pp. 11–22.

Gilmer PJ. 2002. Assessment and students' interest: connecting to learning. In: Wallace J, Louden W, eds. *Dilemmas of Science Teaching: Perspectives on Problems of Practice.* New York, NY: Routledge Publishing. Pp. 48–55.

Gilmer PJ, Rashotte ME. Dec. 1989/Jan. 1990. Marshalling the resources of a large state university for an interdisciplinary 'Science, Technology and Society' course. *Journal of College Science Teaching.* Pp.150–156.

Greenwood MRC, Kovacs-North K. 1999. Science through the looking glass: Winning the battle but losing the war (adapted from the address given by MRC Greenwood at the AAAS Annual Meeting, January 23, 1999). *Science* 286: 2072-2078.

Grinnell F. 1992. *The Scientific Attitude,* 2nd ed. New York, NY: Guilford Press.

Hafferty FW, Franks R. 1994. The hidden curriculum, ethics teaching, and the structure of medical education. *Academic Medicine* 69:861–871.

Heitman E, Salis, P, Bulger, RE 2000. Teaching ethics in biomedical sciences: Effects on moral reasoning skills. Paper presented at the ORI Research Conference on Research Integrity, Washington, DC, November 2000. [Online]. Available: http://ori.dhhs.gov/multimedia/acrobat/papers/heitman.pdf [Accessed March 15, 2002].

Hensel N. 1991. Realizing gender equality in higher education: The need to integrate work/family issues. In: *ASHE-ERIC Higher Education Report No. 2.* Washington, DC: School of Education and Human Development, George Washington University.

Hundert EM. 1996. Characteristics of the informal curriculum and trainees' ethical choices. *Academic Medicine* 71:624–633.

Kegan R. 1982. *The Evolving Self: Problems and Process in Human Development.* Cambridge, MA: Harvard University Press.

King PM, Kitchener KS. 1994. *Developing Reflective Judgment.* San Francisco, CA: Jossey-Bass.

Knowles MS. 1970. *The Modern Practice of Adult Education.* New York, NY: Association Press.

Knox AB. 1978. Helping adults to learn. In: *Yearbook of Adult and Continuing Education.* Chicago, IL: Marquis Academic Media.

Kohlberg L. 1984. *Essays on Moral Development: The Psychology of Moral Development: The Nature and Validity of Moral Stages. Vol. 2.* San Francisco: Harper & Row.

Kovac J. 1995. The Ethical Chemist: Case Studies in Scientific Ethics. Knoxville, TN: Department of Chemistry, University of Tennessee.

Lemke JL. 1995. *Textual Politics: Discourse and Social Dynamics.* London, United Kingdom: Taylor & Francis.

Loucks-Horsley S, Hewson PW, Love N, Stiles KE. 1998. *Designing Professional Development for Teachers of Science and Mathematics Education.* Thousand Oaks, CA: Corwin Press.

Magner DK. 2000, April 28. Critics urge overhaul of Ph.D. training, but sharply disagree on how to do so. *Chronicle of Higher Education.* P. A19.

Mentkowski M. 2000. *Learning That Lasts. Integrating Learning, Development, and Performance in College and Beyond.* San Francisco, CA: Jossey-Bass.

Merriam SB, Caffarella RS. 1999. *Learning in Adulthood,* San Francisco, CA: Jossey-Bass.

NAS (National Academy of Sciences). 1989. *On Being a Scientist.* Washington, DC: National Academy Press.

NAS. 1992. *Responsible Science: Ensuring the Integrity of the Research Process,* Vol. 1. Washington, DC: National Academy Press.

NAS. 1995a. *On Being a Scientist,* 2nd ed. Washington, DC: National Academy Press.

NAS 1995b. *Reshaping the Graduate Education of Scientists and Engineers.* Washington, DC: National Academy Press.

NIH (National Institutes of Health). 1989. *NIH Guide for Grants and Contracts,* Vol. 18, No. 45. Rockville, MD: NIH.

Parker LH, Rennie LJ. 1998. Equitable assessment strategies. In: Fraser B, Tobin KG, eds. *International Handbook of Science Education,* Part 2. Dordrecht, The Netherlands: Kluwer Academic Publishers. Pp. 897–910.

Piaget J. 1932. *The moral judgement of the child.* New York, NY: Free Press.

Posner B, Schmidt W. 1982. What kind of people enter the public and private sectors? An undated comparison of perceptions, stereotypes, and values. *Human Resource Management* 21:35–43.

Posner B, Schmidt W. 1984. Values and the American manager: An update. *California Management Review* 26(3):202–216.

Rest J. 1983. Morality. In: Mussen PH (series ed.) and Flavell J, Markman E. (vol. eds.). *Handbook of Child Psychology,* Vol. 3, *Cognitive Development,* 4th ed. New York, NY: Wiley. Pp. 556–629.

Rest JR, Narváez DF, eds. 1994. *Moral Development in the Professions: Psychology and Applied Ethics.* Hillsdale, NJ: Erlbaum Associates. Pp. 51–70.

Rest J, Bebeau M, Volker J. 1986. An overview of the psychology of morality. In: Rest JR, ed. *Moral Development: Advances in Research and Theory.* New York, NY: Praeger. Pp. 1–27.

Rest J, Thoma SJ, Narváez D, Bebeau MJ. 1997. Alchemy and beyond: Indexing the Defining Issues Test. *Journal of Educational Psychology* 89:498–507.

Rest J, Narváez D, Bebeau MJ, Thoma SJ. 1999. *Postconventional Moral Thinking: A Neo-Kohlbergian Approach.* Hillsdale, NJ: L. Erlbaum Associates.

Rest JR. 1979. *Development in Judging Moral Issues.* Minneapolis, MN: University of Minnesota Press.

Rule JT, Bebeau MJ. 2001. Integrity and mentoring in research: The story of Irwin D. Mandel. *Quintessence International* 32:61–75.

Self DJ, Baldwin DC. 1994. Moral Reasoning in Medicine. In: Rest JR and Narváez, DF, eds. *Moral Development in the Professions: Psychology and Applied Ethics.* Hillsdale, NJ: Erlbaum Associates. Pp. 147–162.

Sigma Xi. 1986. *Honor in Science.* Research Triangle Park, NC: Sigma Xi, the Scientific Research Society.

Strike K, Posner G. 1992. A revisionist theory of conceptual change. In: Duschl R, Hamilton R, eds. *Philosophy of Science, Cognitive Psychology, and Educational Theory and Practice.* Albany, NY: State University of New York. Pp. 211–231.

Tight M. 1996. *Key Concepts in Adult Education and Training.* New York, NY: Routledge Publishing.

Tobin AJ, Dusheck J. 2001. *Asking About Life,* 2nd ed. New York, NY: Harcourt College Publishing.

Tobin K, Tippins D. 1993. Constructivism as a referent for teaching and learning. In: Tobin K., ed. *The Practice of Constructivism in Science Education.* Washington, DC: AAAS Press. Pp. 3–21.

Varmus H. 1995. Statement at the July 13 hearing on the National Academy of Sciences Report *Reshaping the Graduate Education of Scientists and Engineers* of the Subcommittee in Basic Research, Committee on Science, U.S. House of Representatives.

von Glasersfeld E. 1989. Cognition, construction of knowledge, and teaching. *Synthese* 80:121–140.

Walker OC, Churchill GA, Ford NM. 1979. Where do we go from here? Selected conceptual and empirical issues concerning the motivation and performance of the industrial sales force. In Albaum G, Churchill GA, eds. *Critical Issues in Sales Management State of the Art and Future Research Needs.* Eugene, OR: University of Oregon Press. Pp. 10–75.

Widnall SE. 1991. AAAS Presidential Lecture: Voices from the pipeline. *Science* 19:404–419.

Wlodkowski RJ. 1999. *Enhancing Adult Motivation to Learn.* San Francisco, CA: Jossey-Bass.

Zigmond MJ, Fischer, BA. 1995. *A Suggestion of Fraud. In Science Conduct On-Line, Science: Beyond the Printed Page.* [Online]. Available: http://sci.aaas.org/aas-bin/forum-user3.pl [Accessed June 23, 1999].

Zigmond MJ, Bloom FE, Landis SC, Roberts JL, Squire LR, eds. 1999. *Fundamental Neuroscience.* San Diego, CA: Academic Press.

6

Evaluation by Self-Assessment

Various elements of programs intended to enhance the integrity of institutional research were described in Chapter 4, and many of these elements have demonstrated at least a measure of success in some circumstances and by some, often vague benchmarks. The committee concluded, however, that the principal mode for evaluation of the effectiveness of an integrated program should be based on self-assessment and peer review, particularly when undertaken in the context of institutional accreditation. The other elements of an effective program—performance-based assessment, education, and attention to regulatory compliance—are generally components of rigorous institutional self-assessment. Use of self-assessment as a principal tool directly extends, with greater specificity, the first two recommendations of the 1992 National Academy of Sciences (NAS) Panel on Scientific Responsibility and the Conduct of Research (NAS, 1992):

- Individual scientists in cooperation with officials of research institutions should accept formal responsibility for ensuring the integrity of the research process. They should foster an environment, a reward system, and training processes that encourage responsible research practices. (p. 13)
- Scientists and research institutions should integrate into their curricula educational programs that foster faculty and student awareness of concerns related to the integrity of the research process. (p. 13)

The committee endorses the principle of self-assessment as an ante-cedent to formal appraisal of the performances of academic departments and individual faculty members. This chapter discusses in further detail the processes of self-assessment at the levels of both the institution and the research unit (at the level of the department, the research group, and the individual investigator), and it offers some initial steps that might be taken in the application of self-assessment to evaluation of the environment for integrity in research.

SELF-ASSESSMENT AND ACCREDITATION IN HIGHER EDUCATION

State and federal governments mandate accreditation of institutions of higher education as a requirement for the recognition of the degrees they grant, but a process of peer review is used almost exclusively to grant accreditation. Many different accrediting bodies exist, and these are based either on geography or, for professional schools, on the degree granted. In virtually every case, the heart of the accreditation process is self-assessment (Borden and Owens, 2001; Ewell and Lisensky, 1988; Middle States Commission on Higher Education, 2000).

Process of Self-Assessment in Higher Education

Self-assessment begins with instructions from the accrediting body regarding the criteria for evaluation. These instructions generally provide a template for self-assessment that enables the institution to respond to a series of "must" and "should" standards. The issues and questions posed are usually of a general nature so that institutions can present their solu-tions in different ways. These responses are then judged by external re-viewers and provide the basis of an institution's case for accreditation. Because institutions of higher education vary markedly in their histories, cultures, curricula, and human and physical resources, accreditation is not based on presumptions as to particular "right" answers. In fact, within very broad boundaries, institutional diversity is valued and encouraged.

The process of self-assessment in institutions of higher education is lengthy, costly, and difficult (Ewell, 1991; Middle States Commission on Higher Education, 2000). In the process, institutions critically evaluate their strengths and weaknesses and consequently develop new ideas for self-improvement. Continuous quality improvement is the goal. The in-tent is to accomplish this by associating the process of self-assessment with anticipated improvements in the desired output (e.g., better-edu-cated students). Periodic reaccreditation provides a formal process for evaluation of the results. It cannot be accomplished by simple completion of a mandated checklist, although attention to inputs and processes, as well as outcomes, is required.

Determining the effectiveness of any program of continuous quality improvement is largely qualitative, although some measures can be quantified. For example, to evaluate an outcome of creating better-educated students, institutions may look to improvements in scores on examinations taken by their students interested in pursuing advanced degrees (e.g., the Graduate Records Examination and the Medical College Admission Test), or they may monitor the postgraduation careers of their students to determine what kind of postgraduate programs or professional positions they enter and how well they were prepared (National Center for Higher Education Management Systems, 2001). Institutions that have recently completed an accreditation cycle often serve as models for other institutions as the latter prepare for a similar process, thereby encouraging a culture of quality improvement (Ewell and Lisensky, 1988).

The self-assessment process is organized around a set of faculty committees, each responsible for analysis and recommendations concerning an element or program of the institution considered important in determining institutional effectiveness (Middle States Commission on Higher Education, 2000). Both the openness and the accuracy of the reports produced by these committees are furthered by the inclusion of large numbers of faculty in their membership. The reports are assimilated and integrated into a master document, the self-study report. A template provided by the accrediting body may guide both the format and the general content of the report. This self-assessment is submitted to the accrediting body and is then provided to volunteer peer reviewers, who complete their evaluations with a visit to the institution. The reviewers generally interview members of the faculty, administration, and student body, and then prepare a detailed report that addresses perceived strengths, weaknesses, and areas for desirable or necessary improvement. This report constitutes the basis of a recommendation to the accrediting body regarding the continuation of accreditation.

The committee appreciates that in an open society like that of the United States, different people look at many measures of the quality of institutions of higher education and research (e.g., rankings of institutions according to the amount of support for research they receive from the National Institutes of Health, and U.S. *News and World Report* rankings of schools and educational programs). Nevertheless, governments have traditionally relied upon processes of accreditation to define those entities qualified to provide education at a particular level or in a specific field. Despite this reliance, however, the federal government has generally avoided specific involvement in accreditation processes, largely entrusting them to private-sector accrediting agencies, consistent with the nation's tradition of limited government. The effectiveness of this approach to rigorous institutional accreditation is reflected, at least in part, by the fact that the U.S. system of higher education is admired around the world, as well as by the public's continuing general acceptance of self-monitoring as a means of assessing professional quality.

Promotion of Integrity in Research

Promotion of integrity in the research environment is about institutional culture and behavior, as well as the professional performance of individuals. It is about the system in which research is done. Research trainees learn about the culture of science through experience, mentoring, and formal educational processes. Even the best institutional climate and programs for researchers and trainees, however, will not preclude either research misconduct or nonprofessional behavior. Still, an educational environment that makes clear what is expected of scientists and their teams, combined with systems and institutional behaviors and policies that encourage accurate and careful pursuit of scientific ends, can beneficially influence researchers and trainees.

The committee has defined integrity in research more broadly than an absence of research misconduct. Integrity in research embraces the aspirational standards of scientific conduct rather than simply the avoidance of questionable practices. There is a role for education of students, faculty, and staff regarding not only professional behavior but also the common culture of science that as a whole promotes a research environment of high integrity. It can focus on the joy of rightful discovery and recognition by one's peers. Such education should be viewed in its broadest sense, however, occurring not only through formal instruction but also through the institutional atmosphere, policies, and guidelines, as well as through the quality of mentoring (King, 1999; Swazey et al., 1993).

A Role for Institutional Accrediting Bodies

The committee believes that assessment of the effectiveness of institutional efforts to ensure integrity in the research environment can best be accomplished by incorporating evaluations of integrity in research into existing accreditation processes for institutions of higher education. Benefits that flow from the systematic evaluation of institutional behavior and policies associated with a process of self-assessment and accreditation include the following (Ewell and Lisensky, 1988):

- highlighting a need for change without impairing institutional autonomy or uniqueness;

- efficient coordination of data gathering;
- creation of a permanent capacity for analyzing institutional effectiveness; and
- changing of the attitudes of faculty and staff regarding institutional assessment from negative to positive.

Although peer review of the self-assessment process used to ensure integrity in the research environment could be accomplished by establishing a new accrediting body for the purpose the committee believes this is not the preferred approach. A more attractive alternative is for research institutions to work with established accrediting bodies to incorporate research integrity into overall accreditation processes. The processes of established accrediting bodies should be more effective and more cost-efficient than those of a new entity, whose establishment would constitute one more administrative burden, and thus would encourage cynicism. Moreover, creating a separate body for assessment could easily communicate an undesirable message that an environment that enhances integrity in research can readily be distinguished from one that promotes high-quality education and research more generally. The committee believes that the research mission should be considered as a whole. Thus, it seems reasonable that entities charged with accrediting the quality of education at institutions of higher learning that conduct scientific research might also be charged with reviewing the data from the institutions' self-assessment of their climate and procedures for promotion of integrity in research. The committee is aware, however, that adoption of this recommendation must begin with a substantial commitment from the institutions themselves. Accrediting bodies respond to priorities established by schools and universities in determining the issues to be addressed in the process of accreditation. Consequently, if institutional cultures are to be enhanced, then both the call for change and its implementation must come from research institutions. An important next step will be for universities and university associations, working together, to acknowledge the importance of conducting research and education in research in an environment of high integrity. Certainly the strong governmental and public interest in research integrity will prove ample encouragement to support initiating the peer review process,

Universities and research sponsors should urge accrediting bodies, especially those charged with accrediting education programs with a substantial research mission, to include an evaluation of the environment for promotion of integrity in research in the overall processes of accreditation. This objective might first be accomplished through collaboration with specialized accrediting bodies for science-based professions, such as medicine and engineering. Although not always realized (Bayles, 1981), a hallmark of the concept of professionalism and an integral factor in the social contract of professions with society is a presumption of integrity and self-policing (ABIM, 2001; Camenisch, 1996). Schools of engineering have led in explicitly tying this responsibility to accreditation requirements. Recently, the Accreditation Board for Engineering and Technology has established criteria which mandate that graduates demonstrate an understanding of professional and ethical responsibility (Engineering Accreditation Commission, 2000; Rabins, 1998). Additionally, research in schools of medicine is largely supported by the U.S. Department of Health and Human Services, in which the Office of Research Integrity (ORI) resides. Thus, if ORI were to support the development of pilot programs to evaluate this approach, the Liaison Committee on Medical Education, which is charged with accreditation of programs for medical education (LCME, 2001), and the Accreditation Board for Engineering and Technology might be effective places to begin.

The heterogeneity of the accreditation process among major accrediting bodies is acknowledged. For example, the process of self-study leading to accreditation of research universities may be less prescriptive than that for professional schools in regard to the specific areas mandated for evaluation. Universities are commonly allowed substantial leeway and choice in determining the elements examined in their self-assessment process. Accrediting bodies, however, respond to the demands of their client communities. Thus, if institutions of higher education regarded the integrity of research as an element essential for their accreditation, then it is likely that assessments of integrity in research would be incorporated into both the self-study and peer-review phases of accreditation.

It should also be noted that self-assessment of an institution's environment for integrity in research does not depend inextricably upon formal ties to a process for institutional accreditation. The need for independent self-assessment has already been recognized by a number of prominent institutions absent a mandate (Center for Academic Integrity, 2001). Other mechanisms for peer review can be developed, but the committee concludes that explicit and public processes for external peer review help to ensure credibility and public confidence. In institutions where accreditation is not available or cannot for some reason be incorporated into institutional processes of accreditation, other approaches to the provision of external validation should be explored. This is particularly important in the large number of private research institutes and industrial research groups that offer formal programs of education in research at either the predoctoral or the postdoctoral level.

By encouraging accrediting bodies to develop their own designs for review of research integrity, the committee anticipates improvements and a confluence of processes over time as these bodies learn from each other and from feedback from the institutions. Although this will be an evolutionary process, it could begin with, and be facilitated by, such agencies as ORI. Implementation should take place as quickly as possible, given the increasing demands for accountability in the use of public funds.

The committee's focus on the quality of an environment that promotes integrity in research within institutions of higher education is not to imply disinterest in such integrity in other contexts. However, virtually all investigators begin their research careers in a university setting; therefore, the university research group can be considered the crucible for education in research. Additionally, as the principal recipients of public research funds, academic research groups have been the major focus of concern for integrity in research.

Self-assessment as a first step in the accreditation of educational programs is predicated on a judgment that the process can lead to important changes in the environment. Experience with the process in other facets of the institutional milieu supports this notion. Desirable changes have successfully been introduced in many universities, in such areas as personal harassment, teaching about sexuality in medical schools, treatment of minorities, addition of instruction in ambulatory care to medical education, and addition of instruction in ethics to law school and business school curricula. The impetus for these changes has come from within the university community and has occurred as a consequence of government initiatives. A more difficult challenge, but one that warrants a substantial research effort, will be to determine, as evidence of effectiveness, whether long-term changes in behavior have been achieved as a consequence of any interventions intended to promote integrity in research coupled with processes for self-evaluation (Davis et al., 1999; Parochka and Paprockas, 2001).

A Role for Governments

Governments have an appropriate interest in the effectiveness of programs in self-assessment and accreditation, since they rely on such programs in a number of ways. For example, state licensing boards for many professions require graduation from an accredited school as a condition of licensure. They can also influence curricula through their licensing powers. Similarly, only accredited schools of medicine are eligible to receive federal education grants and to participate in federal loan programs. Additionally, federal agencies sometimes review and approve the accrediting bodies themselves. Such a review, based on a set of federally determined standards, becomes a condition for accepting the findings of these bodies in determining eligibility for the commitment of federal funds (Middle States Commission on Higher Education, 2000). Thus, the federal government may oversee accreditation efforts that are based on private-sector self-assessment and peer review, even though government does not participate directly in the review and accreditation process. These government policies, which restrict recognition of graduates and extension of certain educational programs exclusively to accredited schools, are well accepted by the higher education community. The nongovernment coordinating agency for accreditation of postsecondary education is the Council on Higher Education Accreditation (CHEA, 2001). The U.S. Department of Education maintains a listing of recognized regional, national, institutional, and specialized accrediting bodies (DoEd, 2001).

Since the nation's economy and national security, as well as the health of its people, are heavily dependent on continuing reliable research findings, the federal government and the public rightfully place a high priority on integrity in research. The federal government deserves support in its call for more effective strategies to encourage changes in the environment for the promotion of integrity in research and the evaluation of the outcomes of such changes. However, the committee judges that a direct role per se for the federal government in such programs that assess and accredit institutions for their integrity in research is neither necessary nor desirable. The committee believes instead that self-assessment coupled with peer review of the environment for the promotion of integrity in research is more likely to have a positive impact on those programs and cultures.

Funding

Funding is needed in several important areas: to support research aimed at developing new methods for fostering integrity in research (including research on assessing the effectiveness of such approaches), and to support the everyday operation of assessment programs already in place at numerous institutions. The federal government (through the Office of Research Integrity and various research agencies) and private research foundations can play a role in supporting these efforts. One way for them to encourage this process would be to augment grant programs to provide support for research into the enhancement of integrity in research and assessment of the effectiveness of alternative approaches.

In principle, the costs of conducting programs to gauge the integrity of research conducted under federal sponsorship could be covered by the reimbursement of facilities and administrative costs (so-called indirect costs) associated with federal research grants and contracts. However, administrative costs on grants and contracts to educational institutions (but not to other research entities) have been capped for more than a decade at 26 percent of direct costs (Goldman and Williams, 2000; OMB,

2000). When coupled with the additional fact that most research-intensive universities at present have documented administrative costs in excess of 26 percent (personal communication, T. DeCrappeo, Council on Government Relations, March 18, 2002), the consequence is that universities alone bear any additional costs associated with the development or enhancement of programs that evaluate integrity in research. If further encouragement of an environment for integrity in research is truly a priority for research sponsors, then the sponsors should work cooperatively with educational institutions to share in the funding of such programs, particularly if the intent is to develop best practices rather than simply require minimal compliance with applicable regulations and policies.

A Role for Professional and Scientific Societies

Professional and scientific societies have a key role in developing, promoting, and inculcating codes of research ethics within their memberships. The common culture of science within academic departments and professional and scientific societies provides the peer pressure that promotes professionalism. To date, however, the activities of these societies related to these efforts have been limited.

Among their actions, professional and scientific societies should examine the requirement for membership and standing relative to integrity. Are there organizational sanctions that apply to members who have been shown to engage in misconduct, for example? In addition, the organizations should ensure that content relative to the responsible conduct of research is included at their annual meetings, in their journals and other publications, and in other organizational venues for communication.

Well-crafted society codes of research ethics (ACS, 2001; ASBMB, 2001) could be used as guides in the development of both specific objectives and processes for the development of institutional self-assessment programs. Each scientific society should express its standards for the responsible conduct of research and take steps to ensure that its members know these standards. It is important to note that major research universities are homes to a large number of different disciplines. The issues should be regarded as equally important across the entire spectrum of general and discipline-specific scientific societies, including societies involved in the life and earth sciences; social sciences; chemical and physical sciences; and mathematical, computational, and computer sciences.

INDIVIDUAL SELF-ASSESSMENT

Self-assessment for accreditation is an institutional process. It may involve global accreditation of the educational environment and programs or evaluation of particular institutionally imbedded structures, such as programs for evaluation of animal care and use or the protection of human research subjects. A second useful level of self-evaluation is directed at the individual researcher or academic unit as an initial step in periodic appraisals of performance. Many institutions and departments have formal processes for assessment of faculty members and academic leaders (deans and chairs). Faculty members are usually evaluated annually. For deans and chairs the interval is usually longer, and the evaluation serves as a component of in-depth school or departmental reviews.

The academic community accepts these processes on the basis of a presumption that they promote and reflect individual, departmental, and institutional excellence. Department chairs commonly regard annual self-evaluation of faculty members, followed by a formal discussion with the chair, as an important aspect of faculty mentoring. The committee concurs, and it advocates the inclusion of questions in the self-assessment process that evaluate behaviors that promote integrity in research. Faculty members should be asked to evaluate not only their own behaviors as researchers but also interpersonal relations and their research environments. Similarly, as chairs and deans are formally evaluated, questions in self-assessment forms should address the extent to which their leadership within the institution promotes a culture of integrity in research. Responses to such questions could then be incorporated more formally into institutional self-assessment as a component of the institutional accreditation process.

There are few data regarding the effectiveness of individual self-assessment in altering behavior, particularly behavior that is deviant or lacking in integrity, and this is another area in which research is needed. However, individual self-assessment does provide a useful framework for a formalized process of evaluation by institutional superiors. Additionally, such a process may have salutary effects, at least in terms of interpersonal relationships within a laboratory or department, particularly when it includes input from subordinates to the individual being evaluated.

SUMMARY

• Evaluation of the institutional environment for the promotion of integrity in research should be based on processes of self-assessment and peer review.
• Evaluation of integrity in research should be incorporated into existing processes for accreditation of educational and research institutions. Creation of a new entity specifically for accreditation of an environment

that promotes integrity in research is probably unnecessary; it would be costly and likely impose burdens disproportionate to the benefits.

• Assessment of integrity in research for research groups should be a component of regular performance appraisals for faculty and academic leadership.

• Effective self-assessment will require the development and validation of evaluation instruments and measures.

• Federal research agencies and private foundations should support the development of programs to integrate self-assessment of the environment for integrity in research into accreditation processes through interactions with and among stakeholders, and they should fund research into the effectiveness of such programs.

• Federal research sponsors should work with educational institutions to develop funding mechanisms to support programs devoted to promoting the responsible conduct of research.

REFERENCES

ABIM (American Board of Internal Medicine). 2001. *Project Professionalism.* [Online]. Available: http://www.abim.org /pubs/default.htm [Accessed January 22, 2002].

ACS (American Chemical Society). 2001. *Academic Professional Guidelines.* [Online]. Available: http://chemistry.org/portal/servlet/resources/org/chemistry/avercom/display/ContentRetrievalServlet/ACS/ACSContent/careers/apg_Jan01_final.pdf [Accessed November 5, 2001].

ASBMB (American Society for Biochemistry and Molecular Biology). 2001. *ASBMB Code of Ethics.* [Online]. Available: http://www.asbmb.org/ASBMB/ASBMBSiteII.nsf/MenuHomePage/PublicAffairs [Accessed November 5, 2001].

Bayles MD. 1981. *Professional Ethics.* Belmont, CA: Wadsworth Publishing Co.

Borden VMH, Owens JLZ. 2001. *Measuring Quality: Choosing Among Surveys and Other Assessments of College Quality.* Washington, DC: American Council on Education.

Camenisch PE. 1996. The moral foundations of scientific ethics and responsibility. *Journal of Dental Research* 75:825–831.

Center for Academic Integrity. 2001. *Academic Integrity Assessment Guide.* [Online]. Available: http://www.academicintegrity.org/assessGuide.asp [Accessed November 5, 2001].

CHEA (Council on Higher Education Accreditation). 2001. *Good Practice Database.* [Online]. Available: http://www.chea.org/good-practices/index.cfm [Accessed March 18, 2002].

Davis D, O'Brien MAT, Freemantle N, Wolf FM, Mazmanian P, Taylor-Vaisey A. 1999. Impact of formal continuing medical education. *Journal of the American Medical Association* 282:867–873.

DoEd (U.S. Department of Education). 2001. *Overview of Accreditation.* [Online]. Available: http://www.ed.gov/offices/OPE/accreditation [Accessed March 18, 2002].

Engineering Accreditation Commission. 2000. *Criteria for Accrediting Engineering Programs.* [Online]. Available: http://www.abet.org/images/Criteria/2002-03EACCriteria.pdf [Accessed March 18, 2002].

Ewell P. 1991. *Benefits and Costs of Assessment in Higher Education: A Framework for Choice Making.* Boulder, CO: National Center for Higher Education Management Systems.

Ewell PT, Lisensky RP. 1988. *Assessing Institutional Effectiveness.* Washington, DC: Consortium for the Advancement of Private Higher Education.

Goldman CA, Williams T. 2000. *Paying for University Research Facilities and Administration.* Santa Monica, CA: RAND.

King PM. 1999. Why Are College Administrators Reluctant to Teach Ethics? *Synthesis: Law and Policy in Higher Education* 10(4):756-757. Asheville, NC: College Administration Publications, Inc.

LCME (Liaison Committee on Medical Education). 2001. *Accreditation Procedures.* [Online]. Available: http://www.lcme.org/procedur.htm [Accessed March 18, 2002].

Middle States Commission on Higher Education. 2000. *Designs for Excellence: Handbook for Institutional Self-Study,* 7th ed. Philadelphia, PA: Middle States Commission on Higher Education. Pp. 1–4, 13–21.

NAS (National Academy of Sciences). 1992. *Responsible Science: Ensuring the Integrity of the Research Process.* Panel on Scientific Responsibility and the Conduct of Research. Washington, DC: National Academy Press.

National Center for Higher Education Management Systems. 2001. *Comprehensive Alumni Assessment Survey (CAAS).* [Online]. Available: http://www.nchems.org/Surveys/caas.htm [Accessed November 5, 2001].

OMB (Office of Management and Budget). *Circular A-21, Cost Principles for Educational Institutions.* 2000. [Online]. Available: http://www.whitehouse.gov/omb/circulars/a021/a021.html [Accessed March 18, 2002].

Parochka J, Paprockas K. 2001. A continuing medical education lecture and workshop, physician behavior and barriers to change. *Journal of Continuing Education in the Health Professions* 21:110–116.

Rabins MJ. 1998. Teaching engineering ethics to undergraduates: Why? what? how? *Science and Engineering Ethics* 4:291–302.

Swazey JP, Anderson MS, Lewis KS. 1993. Ethical problems in academic research. *American Scientist* 84:542–553.

7

Concluding Remarks and Recommendations

Several overarching conclusions emerged as the Committee on Assessing Integrity in Research Environments addressed the need of the U.S. Department of Health and Human Services (DHHS) to develop means for assessing and tracking the state of integrity in the research environment:

• Attention to issues of integrity in scientific research is very important to the public, scientists, the institutions in which they work, and the scientific enterprise itself.

• No established measures for assessing integrity in the research environment exist.

• Promulgation of and adherence to policies and procedures are necessary, but they are not sufficient means to ensure the responsible conduct of research.

• There is a lack of evidence to definitively support any one way to approach the problem of promoting and evaluating research integrity.

• Education in the responsible conduct of research is critical, but if not done appropriately and in a creative way, education is likely to be of only modest help and may be ineffective.

• Institutional self-assessment is one promising approach to assessing and continually improving integrity in research.

RESEARCH AGENDA

The committee found that existing data are insufficient to enable it to draw definitive conclusions as to which elements of the research environment promote integrity. The elements discussed in Chapter 2 appear to be associated with integrity in research, but the specific contribution of each element remains poorly defined. Empirical studies evaluating the ethical climate before and after implementation of specific policies or practices are lacking; as a consequence, the decision to implement particular programs is often based on anecdotal evidence. True misconduct is rare, and statistics on misconduct are approximate. Thus, looking for a decrease in rates of misconduct is not a viable way to assess the effectiveness of measures implemented to foster integrity in research. In addition, although it is relatively easy to catalog lists of policies and procedures, it is much less straightforward to measure performance and outcomes in the research environment.

Because of the limited empirical data on factors influencing responsible conduct in the scientific environment, the committee drew on more general theory (e.g., theories of organizational behavior, ethical decision making, and adult learning) to formulate the suggestions presented in this report. The findings and conclusions are based on the committee's collective knowledge and experience after its review of the available literature in the science and business arenas as well as its discussions with experts who presented talks at the committee's open meetings.

On the basis of the available information, the committee has described practices that promote the responsible conduct of research (Chapter 2) and has presented a theoretical model (Chapter 3) that contains many of the key components of the research environment and their interactivity. However, this is relatively new territory that needs to be examined with greater precision. Generating specific empirical data on integrity in scientific research is essential to help institutions determine the effectiveness of their efforts to foster a research climate that promotes integrity. Such data will also aid them in the development of better programs and policies in the future.

The request for applications issued by the Office of Research Integrity (ORI) of DHHS on May 2, 2001 (Research on Research Integrity. RFA-NS-02-005), is an important first step toward this goal, as it highlights a variety of potentially productive research topics, as does the ORI website (http://www.ori.dhhs.gov/html/programs/potentialrestopics.asp). The committee believes these topics are best studied in the context of the model presented in Figures 3-1 and 3-2. In addition to the important research questions identified by ORI in its program announcements, the committee identified additional topics that warrant further study.

Methods and Measures

Gaining the methodological expertise needed to carry out research on the relationship between the research environment and integrity in research will require the development and validation of measures, particularly indicators that are observable and quantifiable within the research environment. For example, existing means of conceptualization and measurement of the organizational climate will have to be adapted to the specific context of the assessment of the ethical climate within the research environment.

Furthermore, to measure the outcomes of efforts related to fostering integrity in the research environment, either new instruments must be designed and validated, or existing outcomes and measures (see Appendix B for examples) must be modified and validated in the specific context of the assessment of the ethical climate within the research environment. This development of reliable and valid measures can take considerable time and effort, but it is a necessary first step in a research process leading to a better understanding of the relationship between the research environment and integrity in research. Note that two distinct types of measures should be considered: measures that assess the integrity of the institution with respect to the conduct of research and measures that assess aspects of the integrity of the individual (see Chapter 2 and Appendix B).

Existing methods and measures, examples of which are described in Appendix B, provide models that could be adopted or adapted to evaluate the factors of culture and climate that promote integrity in research. Similarly, Appendix B also provides examples of measures that have successfully been used to assess learning outcomes in professional ethics programs.

Elements of the Research Environment

Research is needed to fully understand the roles of the various elements of the environment that foster the responsible conduct of research. Questions to be considered include the following:

Organizational structure

In what ways do variations in organizational structure (e.g., the size of an institution, the importance of research within the institution, institutional review board composition and procedures, and reward systems) affect the ethical and moral climate and the responsible conduct of research?

Physical structure

Does the physical structure and layout of the research space, or how the space is allocated, affect the ethical and moral climate and the responsible conduct of research? For example, what are the effects of open spaces versus closed spaces for conducting research? What are the effects of various groupings of people within these spaces?

Funding

What is the relationship between the availability of and competition for funding and the responsible conduct of research?

Incentives and rewards

How do existing incentive and reward systems within and outside universities affect the responsible conduct of research? What, if any, aspects of these systems are counterproductive in fostering integrity in research?

Collaboration

How is integrity in research affected by collaborations within and across institutions?

Effectiveness of codes of conduct and honor codes

Do honor codes and professional codes of conduct foster integrity in research? If so, under what conditions do they have an impact?

RECOMMENDATIONS

To facilitate the assessment and promotion of integrity in the research environment, the committee makes several recommendations, which are presented in the sections that follow. In combination, these recommendations are aimed at efforts to foster integrity in research at the individual and institutional levels and to ensure continuous institutional self-assessment and quality improvement.

Future Research

RECOMMENDATION 1: Funding agencies should establish re-

search grant programs to identify, measure, and assess those factors that influence integrity in research.

• The Office of Research Integrity should broaden its current support for research to fund studies that explore new approaches to monitoring and evaluating the integrity of the research environment.

• Federal agencies and foundations that fund extramural research should include in their funding portfolios support for research designed to assess the factors that promote integrity in research across different disciplines and institutions.

• Federal agencies and foundations should fund research designed to assess the relationship between various elements of the research environment and integrity in research; similarities and differences across disciplines and institutions should be determined.

As discussed earlier in this chapter, further research in needed to (1) develop and validate assessment methods and measures and (2) fully understand the roles of the various elements of the research environment in the responsible conduct of research. The results of such research will allow for more effective implementation of the following recommendations.

Institutional Commitment to Integrity

RECOMMENDATION 2: Each research institution should develop and implement a comprehensive program designed to promote integrity in research, using multiple approaches adapted to the specific environments within each institution.

• It is incumbent upon institutions to take a more active role in the development and maintenance of climate and culture within their research environments that promote and support the responsible conduct of research.

• The factors within the research environment that institutions should consider in the development and maintenance of such a culture and climate include, but are not limited to, supportive leadership, appropriate policies and procedures, effective educational programs, and evaluation of any efforts devoted to fostering integrity in research.

• Federal research agencies and private foundations should work with educational institutions to develop funding mechanisms to provide support for programs that promote the responsible conduct of research.

Integrity in research is critical to the progress and acceptance of science. Although a high level of integrity generally characterizes the research community today, lapses in integrity do occur, and some are destructive. It is in the interest of the entire research community that there be sustained, systematic, and explicit efforts to ensure integrity in research. It is important that all institutions have a clear organizational structure and an unambiguous designation of who has the authority and responsibility for research integrity. Institutional leaders should set the tone for their institutions with their own actions. Senior researchers should set an example, not only in their own research practices but also in their willingness to engage in dialogue about ethical questions that arise. Because of the ever-changing nature of science, the research community needs to continuously adapt and improve upon its traditions of responsible behavior, communication, education, and policies with regard to integrity in research.

Federal research agencies and private foundations are appropriate sponsors of grant programs to support research into the development of programs to promote integrity in research and the assessment of the effectiveness of such approaches. In addition to funding the process of development and validation of programs, financial resources are needed for the ongoing implementation of the programs themselves. In principle, costs associated with federally sponsored research could be supported through the indirect costs associated with federal research grants and contracts. However, administrative costs on grants and contracts to educational institutions (but not to other research entities) have been capped, and universities alone now bear the additional costs associated with the development or enhancement of programs that promote or evaluate integrity in research.

Education

RECOMMENDATION 3: Institutions should implement effective educational programs that enhance the responsible conduct of research.

• Educational programs should be built around the development of abilities that give rise to the responsible conduct of research.
• The design of programs should be guided by basic principles of adult learning.
• Integrity in research should be developed within the context of other relevant aspects of an overall research education program, and instruction in the responsible conduct of research should be provided by faculty who are actively engaged in research related to that of the trainees.

Given the large variation in the human contribution to the research organization, the committee believes that it is particularly important for institutions to create an environment in which scientists are able to gain an awareness of the responsible conduct of research as it is defined within today's culture. They need to understand the importance of these standards and expectations, acquire the capacity to resolve ethical dilemmas, and recognize and be able to address conflicting standards of research conduct (see Chapter 5). For lasting change in ethical climate to occur, changes in an institution's curriculum content alone are not sufficient. Attention also needs to focus on how education in the responsible conduct of research is conducted.

The processes that give rise to the responsible conduct of research include the ability to (1) identify the ethical dimensions of situations that arise in the research setting and the laws, regulations, and guidelines that govern one's field (ethical sensitivity); (2) develop defensible rationales for a choice of action (ethical reasoning); (3) integrate the values of one's professional discipline with one's own personal values (identity formation) and appropriately prioritize professional values over personal ones (moral motivation and commitment); and (4) perform with integrity the complex tasks (e.g., communicate ideas and results, obtain funding, teach, and supervise) that are essential to one's career (survival skills).

Education in the responsible conduct of research should (1) be provided within the context of the overall educational program, including as part of mentor-student interactions, the core discipline-specific curriculum, and explicit education in professional skills; (2) take place over an extended period of time—preferably the entire educational program—and include review, practice, and assessment; and (3) involve active learning, including interactions among the instructors and the trainees.

Educational efforts related to the responsible conduct of research should be designed to reach all those involved in scientific research at all levels. Without formal training for existing senior researchers and an instructional program for new researchers, an institution will not be able to develop a consistent message to trainees and students.

Institutional Self-Assessment

RECOMMENDATION 4: Research institutions should evaluate and enhance the integrity of their research environments using a process of self-assessment and external peer review, in an ongoing process that provides input for continuous quality improvement.

• The importance of external peer review of the institution cannot be overemphasized. Such a process will help to ensure the credibility of the review, provide suggestions for improvement of the process, and increase public confidence in the research enterprise.

• Effective self-assessment will require the development and validation of evaluation instruments and measures.

• Assessment of integrity and the factors associated with it (including educational efforts) should occur at all levels within the institution— for example, at the institutional level, the research unit level, and the individual level. At the individual level, assessment of integrity should be an integral part of regular performance appraisals.

• As with any new program, a phase-in or pilot testing period is to be expected, and the assessment and accreditation process should be continually modified as needed based on results of these early actions.

RECOMMENDATION 5: Institutional self-assessment of integrity in research should be part of existing accreditation processes whenever possible.

• Accreditation provides established procedures, including external peer review, that can be modified to incorporate assessments of efforts related to integrity in research within an institution.

• Entities that currently accredit educational programs at institutions where research is conducted would be the bodies to also review the process and the outcome data from the institution's self-assessment of its climate for promotion of integrity in research. These entities include the six regional organizations that accredit institutions of higher education in the United States, as well as the organizations that accredit professional schools or professional educational programs.

• Federal research agencies and private foundations should support efforts to integrate self-assessment of the research environment into existing accreditation processes, and they also should fund research into the effectiveness of such efforts.

Accrediting bodies rely heavily on the process of institutional self-assessment when reviewing an educational institution (Chapter 6). Institutions critically evaluate their strengths and weaknesses and strive for continuous quality improvement.

The committee believes that the research mission should be considered as a whole, and that evaluation of institutional culture for promotion of integrity in research should be an important component of the overall process of accreditation of educational institutions that conduct scientific research. Thus, it seems reasonable that entities charged with accrediting the quality of education at institutions of higher learning that conduct scientific research should be charged with reviewing the process and the

INTEGRITY IN SCIENTIFIC RESEARCH

outcome data from the institution's self-assessment of its climate for promotion of integrity in research.

In institutions where accreditation is not available (e.g., freestanding research institutes) or where this additional mandate cannot be incorporated into existing institutional processes of accreditation, other approaches to ensuring external validation should be explored.

RECOMMENDATION 6: ORI should establish and maintain a public database of institutions that are actively pursuing or employing institutional self-assessment and external peer-review of integrity in research.

• This database should initially include institutions that receive funding for, or are actively engaged in, the development and validation of self-assessment instruments.

A publicly available informational database of ongoing efforts in institutional self-assessment and peer review could serve two purposes. First, the database could serve as a resource for other institutions seeking to develop their own programs, and second, it could serve as an accountability instrument, enabling the public to see which institutions are receiving public funding to develop such programs. ORI, as the federal entity formally charged with developing and implementing activities to promote research integrity as well as being one of the federal agencies that will fund research in this area, is the appropriate locus for this task. ORI would also be a centralized location of the information, which would be preferable to developing multiple databases scattered throughout the professional societies of different disciplines.

CONCLUSION

Integrity in research is essential for maintaining scientific excellence and keeping the public's trust. Research institutions bear the primary burden of promoting and monitoring the responsible conduct of research. They must consistently and effectively provide members of research teams with the resources they need to conduct research responsibly. These resources include leadership and example, training and education, and policies and procedures, as well as tools and support systems. What is expected of individuals should be unambiguous, the consequences of one's conduct should be clear, and anyone needing assistance should have ready access to knowledgeable leaders. Individuals should be able to seek assistance without fear of retribution. Research institutions, accrediting agencies, and public and private organizations that fund research should collaborate to establish and ensure the integrity of the scientific research enterprise.

Appendixes

A

Data Sources and
Literature Review Findings

The Committee on Assessing Integrity in Research Environments explored various data sources in its effort to comprehensively address the task of providing the Office of Research Integrity (ORI) of the U.S. Department of Health and Human Services (DHHS) with a means for tracking the state of integrity in the research environment. In addition to reviewing the professional literature, the committee also reviewed relevant articles and editorials in the popular and scientific press, reviewed federal reports, and examined relevant regulations and guidelines. The committee invited experts to make public presentations, commissioned background papers, and sought additional expert technical assistance from knowledgeable individuals.

LITERATURE REVIEW

Search Terms

The committee began its review by conducting a preliminary literature search. During and after its first meeting, the committee compiled a list of suggested search terms to be used while conducting literature searches (Table A-1). Committee members, Institute of Medicine staff, and the study sponsor suggested terms.

TABLE A-1   Search Terms

Competitive behavior	Research environments
Conflict of interest	Research fraud
Data access	Research integrity (integrity in research)
Data sharing	Research misconduct
Education and research integrity	Research moral and ethical aspect
(integrity in research)	Research norms
Evaluation research	Research productivity
Fabrication	Research standards
Falsification	Responsibility in research
Mentoring	Retraction of publication
Organization of research	Scientific error
Organizational culture	Scientific fraud
Organizational mistakes	Scientific integrity
Peer review	Scientific misconduct
Plagiarism	Selection bias
Professional ethics	University-industry relationships
Public policies/guidelines	Whistle-blowers
Publication bias	White-collar crime
Quality control and research	

Databases

Searches were performed in OVID in the following databases: AGRICOLA, BioethicsLine, Biosis Previews, CSA-Life Science, ERIC, Medline, PsycInfo, Sociological Abstracts, and Wilson/Biological and Agricultural Index (a description of OVID and of each of the databases can be found at http://www.ovid.com/products/databases/index.cfm). Significant overlap was found among the articles identified in the databases. The most comprehensive and useful databases for the committee's purposes were BioethicsLine, Medline, and PsycInfo, as the majority of articles identified in AGRICOLA, Biosis Previews, CSA-Life Science, ERIC, Sociological Abstracts, and Wilson/Biological and Agricultural Index were also listed in one or more of those three databases (BioethicsLine, Medline, and PsycInfo).

Results of Preliminary Literature Search

The initial search of the databases mentioned above, using the keywords listed in Table A-1, yielded more than 16,000 citations. The first round of screening eliminated entries that were not in English, duplicate listings, and duplicate articles in different journals. For the purposes of the committee's task, the entries were narrowed to those published in the past seven years. Note that this exclusion criterion was not inflexible and that some articles and books published before 1996 (such as those of historical interest or published by leaders in the field) were retained in the list. Although the fact that a citation was for a news item, an editorial, or a letter was not a strict exclusionary criterion, most news items, editorials, and letters were not included. By using these criteria, the list was reduced to slightly more than 800 items. In a case-by-case review of the remaining 800 items, articles and books on completely unrelated topics were eliminated. According to the committee's task, articles and books that discussed topics that were unrelated to the research environment were also eliminated. The final list contained 331 items from journals (including primarily articles and reviews, as well as selected editorials, letters, and news items) and 25 books.

The articles retained were published in 132 different journals, encompassing the specialties of dentistry, education, engineering, law, medicine, nature, nursing, nutrition, psychiatry, and research. Eighteen journals had three or more relevant articles (Table A-2). The committee's search revealed a trend similar to that identified by Steneck (2000), in that several journals stand out as leaders in publishing articles on research integrity and the research environment, including *Science*, the *Journal of*

TABLE A-2   Number of Relevant Articles, by Journal

Journal	Number of Relevant Articles
Science	34
Journal of the American Medical Association	31
Academic Medicine	27
Science & Engineering Ethics	22
BMJ (British Medical Journal)	16
Lancet	9
Nature	8
Accountability in Research	5
Journal of Dental Research	5
Professional Ethics	5
Journal of Higher Education	4
Journal of Law, Medicine & Ethics	4
Proceedings of the Society for Experimental Biology & Medicine	4
Annals of Emergency Medicine	3
College Student Journal	3
Critical Reviews in Biomedical Engineering	3
Ethics & Behavior	3
Journal of Medical Ethics	3

TABLE A-3 Number of Relevant Articles, by Category

Category	Number of Citations
Books	25
Codes of ethics	12
Conflict of interest	47
Education	34
Integrity	64
Methodology/evaluation	4
Misconduct	67
Oversight	23
Publications	43
Whistle-blowers	12

the American Medical Association, Academic Medicine, and *Science & Engineering Ethics.* Together, these four sources account for a full one-third of the listed items from journals. (Note that the majority of items from *Science* are news items or editorials.)

The 331 articles were sorted into 10 major categories, as shown in Table A-3. The categories used were codes of ethics; conflict of interest (including funding and intellectual property); education (including mentoring, training, and staff development); integrity (including ethics, morals, and responsible conduct of research); methodology/evaluation (including assessment); misconduct (including fraud); oversight (including monitoring, accreditation, and peer review); publications (including plagiarism and authorship); and whistle-blowers. Books, which tend to cover more than one specific area, are listed separately in Table A-3.

This collection of current literature was available to the committee for its review and analysis over the course of its deliberations.

Additional Literature and Resources

Over the course of the study, current professional literature, the popular and scientific press, and pertinent web sites were continually surveyed for new data and information relevant to the committee's task. The sponsors, invited speakers, and other researchers and professionals also provided literature for the committee's review and consideration.

In addition, Institute of Medicine staff attended professional scientific meetings and symposia during the course of the study to bring back the latest information about integrity in research issues for the committee's review. Among the meetings attended were the Research Conference on Research Integrity, sponsored by ORI; the Medical Research Summit, sponsored by Health Care Compliance Association, Inova Institute of Research and Education, Medical Device Manufacturers Association and the Department of Energy; and, Promoting Responsible Conduct of Research: Policies, Challenges, and Opportunities, a conference sponsored by Public Responsibility in Medicine and Research.

INVITED PRESENTATIONS

Over the course of the study, the committee received and considered information from organizations and individuals representing many different perspectives on research integrity issues.[1] The committee believed that it was important to receive input directly from junior and senior researchers and administrators who routinely address issues of integrity in research in their work (Box A-1).

Speakers and topics were chosen to complement, expand upon, and fill gaps in the committee's own collective expertise. Committee members heard presentations and asked questions to explore fully the data, surrounding issues, and unique perspectives that each speaker provided.

TECHNICAL ASSISTANCE

The committee sought additional expert technical assistance over the course of the study via phone, e-mail, and personal communications with the following individuals: Barbara Brittingham, New England Association of Schools and Colleges; Steven Crow, North Central Association of Colleges and Schools; Beth Fisher, University of Pittsburgh; Alasdair MacIntyre, Notre Dame University; Jean Morse, Middle States Association of Colleges and Schools; George Peterson, Accreditation Board for Engineering and Technology; James Rogers, Southern Association of Colleges and Schools; David Smith, The Poynter Center for the Study of Ethics and American Institutions; David Stevens, Liaison Committee on Medical Education; and Naomi Zigmond, University of Pittsburgh.

COMMISSIONED PAPERS

The committee commissioned several background papers for the committee's use.[2] David H. Guston, associate professor and director, Pro-

---

[1]All written materials presented to the committee were reviewed and considered with respect to the committee's task. This material can be examined by the public at the National Research Council's Public Access Records Office, 2101 Constitution Avenue, NW, Room 171, Washington, DC 20418; telephone: (202) 334-3543.

[2]Commissioned papers may be examined by the public. The public access files are maintained by the National Research Council, which can be reached at (202) 334-3543.

BOX A-1
Invited Presentations

Perspective of the National Science Foundation
    Christine Boesz
    National Science Foundation

Overview of *Responsible Science* (1992)
    Rosemary Chalk
    Institute of Medicine

Convocation on Scientific Conduct (1994) and Planning Workshop for a Guide for
Education in Responsible Science (1997)
    Robin Schoen
    Board on Life Sciences, National Research Council

Overview of Committee on Science, Engineering, and Public Policy (COSEPUP)
Series on Education and Careers in Science
    Deborah Stine
    COSEPUP

Proposed Common Federal Definition of Research Misconduct and Procedures: A
Town Meeting (1999)
    Chris Pascal
    Office of Research Integrity, DHHS

Assessing the Integrity of Publicly Funded Research, a background report prepared for the November 2000 ORI Research Conference on Research Integrity
    Nicholas Steneck
    University of Michigan

Organizations and Integrity: Some Lessons from Managerial Misconduct
    Peter Yeager
    Boston University

Perspectives on the Research Environment
    Howard Schachman
    University of California, Berkeley

---

gram in Public Policy at Rutgers University's E. J. Bloustein School of
Planning and Public Policy, was commissioned to write a review of the
changes with regard to research integrity that have taken place in the 10
years since publication of the National Academy of Sciences' report *Responsible Science: Ensuring the Integrity of the Research Process* (NAS, 1992).
His work provided some of the background material for Chapter 1 and is
the basis for Appendix C.

RCR Training at the National Institutes of Health (NIH)
Joan Schwartz
Assistant Director
Office of Intramural Research, NIH

Scientific Integrity from a Legal Perspective
Barbara Mishkin
Hogan and Hartson

Integrity in the Business Environment
Bart Victor
Vanderbilt University

Perspectives on Scientific Integrity and the Research Environment
Harold Varmus
Memorial Sloan-Kettering Cancer Center

Research Integrity in Graduate Education
Melissa Anderson
University of Minnesota

The Human Side of Research Integrity: A Young Scientist's Perspective
Peter S. Fiske
RAPT Industries

Perspectives on Research Integrity and the Research Environment
Stephanie Bird
Massachusetts Institute of Technology

Perspectives on Research Integrity and the Research Environment
Ruth Fischbach
College of Physicians and Surgeons, Columbia University

Kenneth D. Pimple, director of Teaching Research Ethics Programs at the Poynter Center for the Study of Ethics and American Institutions, was commissioned to write an opinion piece on his personal reflections on the research environment in the United States and to prepare two literature reviews on the following areas of research: (1) empirical assessments of the moral climate in institutions and (2) empirical evaluations of pedagogical approaches to the teaching of research ethics. The results of his searches provided the committee with a comprehensive overview of what literature was available and, equally importantly, what topics were significantly lacking scholarly attention in the literature.

REFERENCES

NAS (National Academy of Sciences). 1992. *Responsible Science: Ensuring the Integrity of the Research Process*, Vol. 1. Washington, DC: National Academy Press.

Steneck NH. 2000. Assessing the integrity of publicly funded research. *Investigating Research Integrity: Proceedings of the First ORI Research Conference on Research Integrity, November 2000.* [Online] Available: http://ori.dhhs.gov/html/publications/rcri.html [Accessed March 14, 2002].

B

Outcome Measures for Assessing Integrity in the Research Environment

This appendix to the report describes outcome measures and models for the development of outcome measures that could be used or adapted for use by institutions and educators who wish to assess integrity in the research environment. These measures can be applied to assessments of individuals or institutions by processes recommended in this report.

The appendix describes two kinds of outcome measures. First, it describes measures that have been used to assess the moral climate of an institution. Although measures have not been developed specifically for assessment of the climate of integrity in the research institution, measures and methods that could be adapted for use by research institutions have been developed in other settings.

Second, the appendix describes measures that have been used to assess aspects of integrity of the individual. The goal is to recommend measures that could be used (or adapted) by researchers or institutions interested in assessing outcomes of educational efforts to promote the development of integrity in research in trainees. The emphasis will be on outcome measures that are theoretically grounded, that are at least indirect measures of behavior, and that either have been effectively used or have good potential for linking the development of aspects of integrity (e.g., ethical sensitivity, moral reasoning and judgment, and identity formation) to institutional effectiveness. In cases in which a recommended measure cannot be used exactly as designed, the criterion for determination of inclusion in this review is whether the method of assessment has been sufficiently well validated—even if it is in a setting other than research—to warrant adaptation to the research environment.

In summary, measures that meet the following criteria are included: (1) they are theoretically well grounded in a model of morality that demonstrates the relationship between aspects of integrity and behavior; (2) they meet or exceed the minimal criteria for validity and reliability; (3) they have been successfully used to assess learning outcomes for adults either in research ethics programs or in professional ethics programs; (4) they have been used effectively to assess institutional effectiveness in promoting one or more aspects of integrity; and (5) the method of measurement is appropriate for assessment of an aspect of integrity in the research environment, even though the content of the measure may be specific to another discipline. Note that this discussion does not include measures or tests that assess content knowledge of the rules related to the conduct of research, measures that assess perceptions of the integrity of others (e.g., survey instruments designed for the Acadia Institute study), or measures designed to assess the norms of scientists with respect to misconduct and questionable research practices (Bebeau and Davis, 1996; Korenman et al., 1998). The latter might serve as a resource for the development of items for use in a survey of the moral climate of an institution or for items for assessment of role concept development.

METHODS AND MEASURES FOR ASSESSING INTEGRITY IN THE RESEARCH ENVIRONMENT

Two bodies of literature contribute to the understanding of moral climate and its importance for the assessment of integrity in the research environment. The first is the literature on individual moral development, indicating that individual characteristics are not sufficient as an explanation for ethical behavior. Thus, efforts to influence behavior by focusing on the development of abilities related to decision making may be necessary, but not sufficient, to affect integrity in the research environment. The second is the literature on organizational culture and climate that highlights the different kinds of cultures that may be operating in the environment. There is a growing belief that organizations are social actors responsible for the ethical or unethical behaviors of their employees. In fact, corporations (Bowen and Power, 1993) have been held responsible under the law for acts of malfeasance and misfeasance engaged in by employees, sometimes even when the acts of those employees were beyond the scope of their employment. Such instances prompted scholars in the field of organizational development to turn their attention to the assessment of moral climate and to an analysis of the effects of moral climate on decision making.

Individual Development and Its Relationship to Collective Norms

In the early 1980s, developmental psychologists working in correctional facilities and high schools introduced the concept of a "moral atmosphere" or "just community" to explain the social context that shaped collective norms, which seemed either to inhibit or to override the influence of individual moral development on behavior. To measure moral atmosphere, researchers (Higgins et al., 1984; Power, 1980; Power et al., 1989) presented students with dilemmas likely to occur in their environment. For example, in a high school setting, researchers might present situations involving someone who cheated on an exam or someone who was rude to others. The researcher elicits judgments of responsibility (e.g., What do you think _____ should do? Why?) and judgments of practicality (e.g., What would you do? Why?). These were contrasted with perceptions of the collective norms (What would most others in your school do in this situation? Why would they do that?). Through interviews, researchers were able to identify collective norms and establish whether the norm emerged from within the group or was stipulated by authority external to the group. Then, the degree to which the norm met moral standards and the degree to which individuals were committed to each norm were assessed.

By use of this strategy, it was possible to detect groups with strong, but morally defective, collective norms.[1] Furthermore, researchers were able to show that groups develop collective norms that belong only to the group. When prosocial collective norms defined what was expected of group members as group members, individuals tended to conform to group norms even when their competence in moral decision making was not well developed. However, when the collective norms did not encourage prosocial behavior,[2] individuals with higher levels of competence in moral development felt alienated and discouraged from engaging in actions consistent with their level of competence. Higgins and colleagues (1984) concluded that practical moral action is not simply a product of an individual's moral competence but is a product of the interaction between his or her competence and the moral features of the situation.

Melissa Anderson, in a National Science Foundation-funded longitudinal study of doctoral students' acquisition of the concepts of science and its norms, uses interview questions similar to those used to elicit

---

[1]Examples of groups with morally defective collective norms might include repressive totalitarian states, fanatical cults, violent gangs, and organized crime.

[2]Psychologists use the term *prosocial behaviors* to distinguish behaviors that are clearly beneficial to another and support societal or communal norms from behaviors that may be norm or rule based (as in a teen-age gang or criminal group) but support the self, or hurt others. A prosocial behavior is not necessarily selfless.

implicit norms that shape behavior in the studies cited above. Anderson describes the interview questions as follows: "A series of questions ask students to consider and comment on the relationship between academic norms and behavior (Do you see any conflicts between what people think or say you should do and the way work is actually done?), between their own perspectives and behavior (Do you see people around here acting contrary to your advice [to doctoral students on how to avoid serious mistakes]?) and between their own normative perspectives and academic norms (Are there any ideas or rules about how you should do your work that you don't agree with?)" (Anderson, 2001, p. 2). Narrative accounts are then analyzed in terms of the contrasts presented above. At a conference sponsored by Office of Research Integrity (ORI), U.S. Department of Health and Human Services (DHHS), in 2001, Anderson reported findings from an analysis of interviews with 30 first-year doctoral students. (See Chapter 5 for a further discussion of the initial findings and their relationship to education in the responsible conduct of research.)

Organizational Literature

Building on the early work on moral atmosphere, which attempted to define collective norms operating in the environment, Cullen and colleagues argued, "that corporations, like individuals, have their own sets of ethics that help define their characters. And just as personal ethics guide what an individual will do when faced with moral dilemmas, corporate ethics guide what an organization will do when faced with issues of conflicting values" (Cullen et al., 1989, p. 50). Ethical climates were conceptualized as general and pervasive characteristics of organizations that affect a broad range of decisions. In the organizational literature, work climate is defined as "perceptions that are psychologically meaningful moral descriptions that people agree characterize a system's practices and procedures" (Cullen et al., 1993, p. 180).

In contrast to the interview strategy, which, although labor intensive, has the advantage of gauging individual concepts of responsibility as well as perceptions of the group norms, Cullen and colleagues (1993) developed and validated a 36-item questionnaire, the Ethical Climate Questionnaire, to assess perceptions of the norms operating within an organization. Examples of items used to assess climate are as follows:

1. In this company, people are mostly out for themselves.
2. The major responsibility for people in this company is to consider efficiency first.

3. In this company, people are expected to follow their own personal and moral beliefs.

4. People are expected to do anything to further the company's interests.

5. In this company, people look out for each other's good.

6. There is no room for one's own personal morals or ethics in this company.

7. It is very important to follow strictly the company's rules and procedures here.

8. Work is considered substandard only when it hurts the company's interests.

9. Each person in this company decides for himself what is right and wrong.

10. In this company, people protect their own interest above other considerations.

11. The most important consideration in this company is each person's sense of right and wrong.

12. The most important concern is the good of all the people in the company.

13. The first consideration is whether a decision violates any law.

14. People are expected to comply with the law and professional standards over and above other considerations.

15. Everyone is expected to stick by company rules and procedures.

Responses to the questionnaire confirm the multidimensional nature of ethical climate and substantiate the existence of a number of hypothesized ethical climates. Victor and Cullen's (1988) measure is well validated, and their studies confirm that ethical climates are perceived at the psychological level and that individuals within organizations are able to describe the moral atmosphere that prevails in their work units. The kinds of moral climates that prevail differ dramatically among organizations. Furthermore, there appears to be variance in the ethical climate within organizations by position, tenure, and work group membership. The authors argue that ethical climates, although relatively enduring, are not static. A careful assessment of the climate enables an organization to reflect on its policies and practices and institute reforms.

Examples of efforts to evaluate the organizational climate in settings that seem relevant to the research environment follow. As useful as these illustrations are for showing how an organization might assess its moral and ethical climate, it is still up to the institution to implement changes and then to reassess the climate to determine whether the improvements have occurred.

Examples of Climate Assessments Conducted in Related Fields

U.S. Office of Government Ethics

In 1999, the U.S. Office of Government Ethics (OGE) hired a consulting firm to assess the effectiveness of the executive branch ethics program and to assess the ethical culture of the executive branch from the employees' perspective (OGE, 2000). The objective of the executive branch ethics program is to prevent conflicts of interest and misconduct that undermine the public's trust in government. The study assessed employee perceptions of the ethical culture in the executive branch and enabled OGE to make specific decisions regarding the ethics training programs for executive branch employees; the effectiveness of communication regarding the purpose, goals, and objectives of the ethics program; and the extent to which the program helped employees avoid at-risk situations. Because the study was a first attempt to assess the ethical climate of the executive branch the study focused on overall awareness rather than an analysis of the climate within individual executive branch agencies.

The OGE survey was based on the IntraSight Assessment, an assessment tool developed by Arthur Andersen researchers and academic researchers in the fields of business ethics and organizational behavior. Whereas the full report claims that the measure is statistically reliable and valid, a summary of validity and reliability data on the measures was not provided. The IntraSight Assessment examines the impact of an organization's ethics program by assessing employees' perceptions of observed unethical or illegal behaviors and several desirable outcomes of ethics efforts. The IntraSight Assessment examines program elements and cultural factors that, in the original study, had the greatest relationship with desirable outcomes. By providing a measure of outcomes and a measure of the related factors, the IntraSight Assessment provides direction for improving outcomes by addressing the factors most highly related to the desired outcomes. The assessment process provided data that OGE could use for continuous quality improvement. One might expect that future efforts at quality assessment would focus on evaluation of the effectiveness of ethics programs within agencies.

Academic Integrity Assessment

The Center for Academic Integrity at Duke University developed a process and measures that assist institutions of higher learning with assessing the extent to which the climate on their campuses promotes academic integrity (Burnett et al., 1998).

The process begins with the appointment of a campus committee charged with evaluating the state of academic integrity on campus and, after a data collection process, drawing conclusions and making recommendations for ways that programs that have been charged with ensuring academic integrity can improve. The committee assembles background information about the policies and disciplinary procedures (including information and statistics about sanctions that have been imposed); collects descriptions of the educational programs and activities that inform students, faculty, and administrators about academic integrity on campus; conducts focus groups for administrators; and facilitates the collection of data on perceptions of the moral climate from students and faculty.

The center conducts surveys using the Student Academic Integrity Survey and the Faculty Academic Integrity Survey designed by Donald McCabe. According to the developer, the surveys can be modified to address specific content issues that may be unique to the institution and to address objectives defined by the committee. The survey has been used in several studies, but the guide to the survey provides no references to the psychometric properties of the survey. A recent communication (January 2002) with the test developer confirmed that there are no published data on the validity of the measure. The developer does periodically check its reliability, and it would be possible for the developer to make the data available. Included in the guide are criteria for review of an institution's policies and disciplinary procedures and outcomes. The center analyzes the data collected by the surveys, as well as comparison data from national samples for the committee's use in examining the results. The committee's final task is to draw conclusions and make recommendations for ways in which the institution's academic integrity programs can be improved.

Additional Examples

The U.S. Army uses the Ethical Climate Assessment Survey and the Framework for Establishing/Changing Ethical Climate as part of leadership development for members of the U.S. military (U.S. Army, 2001). Leaders are directed to periodically assess their unit's ethical climate and take appropriate actions to maintain the high ethical standards expected of all organizations that are part of the U.S Army. According to information from the web site (U.S. Army, 2001), an ethical climate is one in which "stated Army values are routinely articulated, supported, practiced and respected." An organization's climate is determined by "the individual character of unit members, the policies and practices within the organization, the actions of unit leaders, and environmental and mission factors." ECAS is a self-administered questionnaire that leaders use to assess how the leader perceives his or her unit and leader actions. Col. George

Forsythe (personal communication, United States Military Academy, January 2002) indicated that although the Army has used the measure extensively, studies of the validity of the measure have not been systematically conducted.

The National Center for Education Statistics of the U.S. Department of Education compiled the responses of teachers in private and public elementary and secondary schools to an ethical climate survey. The 27-item questionnaire is intended for use by individual schools to assess the organization's ethical culture.

Summary

It is apparent from the number of measures of moral climate that have been developed that scholars, at least scholars in organizational development, accept the notion that institutions differ in the kinds of moral and ethical climates that prevail and that the moral and ethical climate of an institution can influence a broad range of outcomes for which a given institution may be held accountable. There also appears to be a belief that institutions have a responsibility to assess the moral and ethical climate that prevails, to reflect on the policies and practices that contribute to that climate, to make appropriate adjustments, and to reassess their moral and ethical climates. It is also apparent that whereas a number of measures have been developed to document the prevailing moral and ethical climate, with the exception of the measure designed by Victor and Cullen (1988), little attention has been given to establishing that the data collected by such surveys provide an accurate and reliable picture of the prevailing moral and ethical climate. As easy as it may be to adapt items from existing measures to develop a climate survey to be used in research institutions, it is incumbent upon the research community to establish the validity, reliability, and usefulness of such measures.

METHODS AND MEASURES FOR ASSESSING INTEGRITY OF THE INDIVIDUAL

This section provides descriptions of measures or methods used to assess aspects of the moral integrity of the individual. Included are measures of general abilities that are developmental and that are linked to ethical behavior (Bebeau et al., 1999). Measures that assess aspects of the Four-Component Model of Morality of Rest (1983)[3] are described and are

---

[3]See Chapter 5, Box 5-1, for an operational definition of each of the components of morality.

classified under the following headings: ethical sensitivity, ethical reasoning and judgment, identity formation, and ethical implementation. In most cases, the measures described are profession specific, in that the content of the measure would not be appropriate for the assessment of integrity in research. Nonetheless, the competence being assessed is an ability that is relevant to the integrity of the researcher. If the content of the test is adapted, as has been the case in many of the examples cited below, the measurement strategy should be as effective for assessments of important learning outcomes in the research setting as it has been for assessments of important learning outcomes in other professional settings.

Descriptions of some assessment strategies that rely on Rest's Four-Component Model of Morality (1983) for their theoretical grounding and that seem promising for application to research ethics follow.

Ethical Sensitivity

Performance-based methods for assessment of ethical sensitivity were first developed in dentistry (Bebeau et al., 1985), and the most extensive work on the validity of the method has been conducted with the Dental Ethical Sensitivity Test (Forms A and B). (See Rest et al. [1986] and Bebeau [1994, 2001] for summaries of the validation studies.) The general strategies for ethical sensitivity assessment have been applied in other professional settings: counselor education (Brabeck and Weisgerber, 1989; Volker, 1984); computer users (Liebowitz, 1990); undergraduate education (McNeel, 1990; Mentkowski and Loacker, 1985); geriatric dentistry (Ernest, 1990); social work (Fleck-Henderson, 1995); journalism (Lind, 1997); and school personnel, including administrators, teachers, and school psychologists (Brabeck et al., 2000).

An ethical sensitivity test (Bebeau and Rest, 1990; Ernest, 1990) places students in real-life situations in which they witness an interaction on either videotape or audiotape. The interaction replicates professional interactions and provides clues to a professional ethical dilemma. For example, the Racial Ethical Sensitivity Test (Brabeck, 1998) consists of five videotaped scenarios that portray acts of intolerance exhibited by professionals in school settings. Each scenario includes from five to nine acts of racial and gender intolerance that violate one or more of the common principles specified in ethical codes of school-based professions. Distinct from the cases typically used in ethics courses, the information is not predigested or interpreted. At a point in the presentation, the student is asked to take on the role of the professional in the situation and respond (on an audiotape) as though he or she were that person. Following his or her response to a patient, client, or colleague, the student answers a number of probe questions that ask why he or she said what was said; how he or she expects the patient, client, or colleague to respond; what he or she thinks should be done in like situations, and so on. Using established (by an interdisciplinary team that includes practitioners) and well-validated criteria, judges rate the extent to which the student adequately interprets the significant issues and professional responsibilities presented in the situation.

Studies assessing the ethical sensitivities of both professionals in training and professionals in practice (Bebeau, 2001; Bebeau and Brabeck, 1987; Bebeau et al., 1985; Fleck-Henderson, 1995) indicate considerable variability among professionals in terms of their sensitivities to the ethical issues they may encounter. Thus, completion of professional training does not ensure development of sensitivity to professional issues.

Studies also show, however, that ethical sensitivity can be improved with instruction (Bebeau and Brabeck, 1987; Leibowitz, 1990; Mentkowski and Loacker, 1985; Sirin et al., submitted for publication). Furthermore, studies show that ethical sensitivity is distinct from the ability to reason (the second component of the Four-Component Model of Morality of Rest [1983]) about what ought to be done in a situation (Bebeau and Brabeck, 1987; Bebeau et al., 1985; Brabeck et al., 2000). Consequently, one cannot assume that education that focuses on ethical reasoning will transfer the ethical reasoning ability to the interpretive process.

Because the assessment process is relatively expensive, requiring transcription of a semi structured interview and scoring by trained raters, measures of ethical sensitivity have typically been used in research studies. Recently, however, Brabeck and Sirin (2001) produced a computerized version of the Racial Ethical Sensitivity Test (REST-CD), intended to make their test more efficient. A subsequent study (Sirin et al., submitted for publication) concluded that the more efficient assessment process provides a reliable and valid measure of ethical sensitivity to instances of racial and gender intolerance.

The modified ethical sensitivity assessment strategy of Brabeck and colleagues seems ideal for assessment of sensitivity to the cultural, interpersonal, and value conflicts that arise between parties (e.g., mentors and students, collaborators, or administrators and researchers) in the research setting. Notice, however, that in addition to assessing the professional's attention to behaviors of the person, the cases assess knowledge of the rules, regulations, and codes of ethics in the context in which they are used. Tests that assess the application of knowledge in context usually provide better assurances of knowledge acquisition. The cases developed by ethical sensitivity researchers are not unlike the dialogue cases.

Ethical Reasoning and Judgment

Assessing Written Essays

Perhaps the most familiar approach to measuring ethical reasoning and judgment is the analysis of written arguments, typically conducted by faculty who teach philosophy or professional ethics (Howe, 1982). In dentistry (Bebeau, 1994) and nursing (McAlpine et al., 1997), for example, researchers have demonstrated that essays can be reliably assessed and that instruction is effective in promoting the ability to develop well-written essays that meet criteria that are specified in advance of instruction. Such methods lack practicality for the assessment of competence in reasoning as a function of an institution's efforts to promote reasoning about dilemmas in integrity in research, as they are labor intensive and require considerable expertise in philosophy or ethics. However, assessment of written essays is a particularly effective way to promote learning, especially if it is accompanied by clearly stated criteria, frequent opportunities for practice, and feedback (Bebeau, 1994).

These methods have been applied to integrity in research with various degrees of success. For example, Stern and Elliott (1997) describe the challenges in establishing interrater reliability and the lack of a measurable effect if the criteria used to judge moral arguments are not presented as part of the instructional program. Recognizing both the need to teach the criteria used to make judgments about the adequacy of moral arguments and the need to be able to reliably apply the criteria to the evaluation of arguments developed by students, the Poynter Center developed and validated a set of cases and criteria for the assessment of moral reasoning in scientific research.

Moral Reasoning in Scientific Research: Cases for Teaching and Assessment (Bebeau et al., 1995) is an 80-page booklet that features six one- to two-page case studies, as well as extensive information on how to use the case studies and a discussion of the theoretical underpinnings of the approach. In addition to notes that provide the instructor with guidance on leading case discussions, the booklet includes a handout for students that details the criteria used to judge the adequacies of moral arguments. As its title implies, *Moral Reasoning in Scientific Research* is designed to facilitate improvements in moral reasoning skills, as well as to facilitate assessments of such improvements. Evidence of the effectiveness of the techniques for facilitating reasoning and the validity of the assessment are described and referenced in the booklet. Ken Pimple, a coauthor on the project, recently converted the booklet to PDF format and made it available via the Poynter Center's World Wide Web site (Bebeau et al., 1995).

Objective Measures of Moral Reasoning and Judgment

Researchers have developed objective measures of moral reasoning and judgment.[4] The most widely used test that may have the most potential for the assessment of institutional effectiveness in research settings, the Defining Issues Test (DIT) (Rest, 1979; Rest et al., 1999a), has a long validation history and is a well-established measure of student learning outcomes. A large body of literature (Mentkowski, 2000; Pascarella and Terenzini, 1991) has addressed the influence of institutions of higher education on the development of critical thinking, moral development, identity formation, and so on. Of all of the measures that have been designed to show the impact of higher education on important learning outcomes, DIT stands out as one of the best indicators of learning outcomes that can be linked back to institutional effectiveness.

The Defining Issues Test   Developed by the late James Rest (Rest, 1979; Rest et al., 1997, 1999a), DIT is a paper-and-pencil measure of moral judgment based on Lawrence Kohlberg's (1984) pioneering work on the development of moral judgment over the life span. DIT measures the reasoning strategies (moral schemas) that an individual uses when confronted with complex moral problems and the consistency between reasoning and judgment. The test presents six moral dilemmas that cannot be fairly resolved by applying existing norms, rules, or laws. Respondents rate and rank arguments (12 for each problem) that they considered important in coming to a decision about what they would do. The arguments reflect the conceptually distinct reasoning strategies (schemas) that people use to justify their actions. The scores reflect the proportion of times that a person prefers each strategy. The most widely used score, the P Index (where P is for postconventional thinking), describes the proportion of times that a respondent selects arguments that appeal to moral ideals. Research indicates that mature thinkers appeal to moral ideals much more frequently than immature thinkers do. Mature thinkers (e.g., ethicists and thoughtful professionals) attempt to work out what ought to be done in circumstances in which there is a conflict of rights, interests, or obligations. They make modifications to existing rules, laws, or codes of ethics to accommodate the new moral problem that has arisen. Because professionals are often required to apply ethical principles or ideals to new problems that emerge in their professions, this skill is necessary for effective moral func-

---

[4]In addition to the Moral Judgment Interview of Colby and colleagues (1987), Gibbs and colleagues (1992) designed the Sociomoral Reflection Measure, suitable for the assessment of reasoning for children and adolescents, and Lind and Wakenhut (1985) designed the Moral Judgment Test, which has mainly been used in Germany. Each of these measures has been validated and has advantages.

tioning. Research indicates that there is a strong relationship between the P Index and prosocial moral action.

In addition to the P Index, the test also determines the proportion of times that an individual selects arguments based on two other problem-solving strategies: the PI Index (where PI represents personal interests) describes the proportion of times that a respondent selects arguments that appeal to personal interests and loyalty to friends and family, even when doing so compromises the interests of persons outside one's immediate circle of friends, and the MN Index (where MN represents maintaining norms) describes the proportion of times that a respondent selects arguments that appeal to the maintenance of law and order, irrespective of whether applying the law to the dilemma presented results in an injustice. In addition to the three main indices, the program calculates two information-processing indices: the U Index (where U represents utilizer), whose score ranges from -1.0 to $+1.0$ and which describes the degree of consistency between reasoning and judgment (persons whose reasoning and judgments are reasonably consistent achieve scores of 0.4 or above), and the N2 Index, which takes into account how well the respondent discriminates among the various arguments and which is often a better indicator of change than the P Index. If the N2 Index score is higher than the P Index score, it indicates that the respondent is better able to discriminate among arguments than to recognize postconventional arguments.

The validity of DIT has been assessed in terms of seven criteria (Rest et al., 1999a):

1. Differentiation of various age and education groups. Studies show that 30 to 50 percent of the variance of DIT scores is attributable to level of education.

2. Longitudinal gains. A 10-year longitudinal study of men and women, college attendees, and subjects not in college and from diverse walks of life show gains in DIT scores over the 10-year period; a review of a dozen studies of first-year to senior college students ($N > 500$) show effect sizes of 0.80, making gains in DIT scores one of the most dramatic effects of college.

3. Relation to cognitive capacity measures. DIT is significantly related to cognitive capacity measures of moral comprehension ($r = 0.60s$), recall and reconstruction of postconventional moral arguments, Kohlberg's (1984) interview measure, and (to a lesser degree) other measures of cognitive development.

4. Sensitivity to moral education interventions. DIT is sensitive to moral education interventions. One review of more than 50 intervention studies reports an effect size for dilemma discussion interventions of 0.40

(moderate gains), whereas the effect size for comparison groups was only 0.09 (little gain).

5. Linkage to many prosocial behaviors and to desired professional decision making. DIT is significantly linked to many prosocial behaviors and to desired professional decision making. One review reports that the links for 37 of 47 measures were statistically significant.

6. Linkage to political attitudes and political choices. DIT is significantly linked to political attitudes and political choices. In a review of several dozen correlates of political attitude, DIT typically correlates with r values in the range of 0.40 to 0.60. When coupled with measures of cultural ideology, the combination predicts up to two-thirds of the variance of controversial public policy issues (such as abortion, religion in public schools, the roles of women, the rights of accused individuals, the rights of homosexuals, and free speech issues).

7. Reliability is good. The Cronbach alpha value[5] is in the upper 0.70s to low 0.80s. The test-retest reliability of DIT is stable.

Furthermore, DIT shows discriminant validity from verbal ability-general intelligence and from conservative-liberal political attitudes; that is, the information in a DIT score predicts the seven validity criteria above and beyond that accounted for by verbal ability or political attitude. DIT is equally valid for males and females.

DIT-2 (Rest et al., 1999b) is an updated version of the original DIT (DIT-1) devised 25 years ago. Compared with DIT-1, DIT-2 not only has stories that are not dated but is also a shorter test, has clearer instructions, and retains more subjects through subject reliability checks. In addition, in studies conducted so far, the validity of the test is not sacrificed because it is a shorter test. If anything, it improves on validity. The correlation of the results of DIT-1 with those of DIT-2 is 0.78, approaching the test-retest reliability of DIT-1 with itself.

Using DIT to Assess Educational Effects   Because DIT has been used to assess the effects of interventions in professional ethics and research ethics (Heitman et al., 2000), a brief summary of findings is included here.

---

[5]Cronbach alpha (Cronbach, 1951) provides an estimate of the internal consistency of the test. Because ranking data are used to calculate the P index and the N2 index, the individual items would not be the appropriate unit of analysis for determining internal consistency reliability. Further, ranking data are ipsative; that is, if one item is ranked in first place, then no other item can be ranked in first place. Therefore, the unit of internal reliability is on the story level, not the item level, and Cronbach alpha is the appropriate strategy for estimating internal consistency. Calculated across six stories for DIT1, the estimates are 0.76, for the five story DIT2 0.81, which is somewhat lower than the estimate of 0.90 if calculated across all 11 stories for the two forms of the test (Rest et al., 1997).

Typically, researchers have reported scores in terms of the P Index score (the proportion of items selected that appeal to postconventional moral frameworks for decision making). The average adult selects postconventional moral arguments about 40 percent of the time, the average Ph.D. candidate in moral philosophy or political science does so about 65.2 percent of the time, the average graduate student does so 53.5 percent of the time, the average college graduate does so 42 percent of the time, and the average high school student does so 31.8 percent of the time (Rest et al., 1999b).

Similar to college graduates, Heitman and colleagues' (2000) sample of 280 graduate students in a research ethics course achieved a mean score of 43.9 (standard deviation [SD], 13.1). In contrast, a sample of 14 scientists (from a variety of disciplines) who completed DIT while in attendance at a summer institute on the teaching of research ethics achieved a mean score of 53 (SD, 13), comparable to the mean and variance for graduate students. What is important about this data set is that the variability among those interested in teaching research ethics is comparable to the variability observed among students and professionals like physicians and dentists. In other words, one cannot assume the development of postconventional thinking on the basis of one's achievement as a scientist.

Furthermore, a recent analysis of DIT profiles for entering professional students (i.e., the proportion of arguments selected with a personal interest, maintaining norms, and postconventional moral framework) indicates that fully 47 percent of a sample of 222 first-year students were in a "transitional status" of developmental change in their mode of thinking (Bebeau, 2001). In other words, their DIT profiles indicated that they were not distinguishing less adequate from more adequate moral arguments as well as students who had completed their ethics program were. As a consequence of this recent observation and a recent meta-analysis of the effects of interventions on moral judgment development (Yeap, 1999), Bebeau (2001) recommends that researchers studying the effects on an intervention conduct a profile analysis rather than rely only on the P Index as a measure of change.

Whereas progress in moral judgment is developmental and development proceeds as long as an individual is in an environment that stimulates moral thinking, gains in moral judgment are typically not found to be associated with professional education programs (e.g., veterinary medicine, medicine, dentistry, and accounting programs) unless the program has a specially designed ethics curriculum (Rest and Narváez, 1994). Furthermore, for some students (Bebeau and Thoma, 1994) and some professions (Ponemon and Gabhart, 1994), educational programs actually seem to inhibit growth in terms of gaining moral judgment. For example, Ponemon and Gabhart speculate that the heavy emphasis placed on learning and applying regulatory codes in the education of accountants may inadvertently promote a maintaining norms moral framework that inhibits the development of the advanced moral frameworks needed to reason through new moral issues. Such findings reinforce the importance of the use of outcome measures to assess institutional effectiveness in promoting the development of reasoning ability.

Development of a Prototype Intermediate Concept Measure Tests like DIT are valuable for assessment of a general reasoning ability that is a critical element of professional ethical development, but they may not be sensitive to the specific concepts taught in a professional ethics course— or, indeed, in a research ethics course. Referring to teacher education, Strike points out: "It is no doubt desirable that teachers acquire sophisticated and abstract principles of moral reasoning [as measured by DIT]. . . . But a teacher who has a good grasp of abstract moral principles may nevertheless lack an adequate grasp of specific moral concepts, such as due process" (Strike, 1982, p. 213). The question (for educators) is often whether to teach specifically to the codes or policy manuals or to teach concepts particular to a discipline: informed consent, intellectual property, conflict of interest, and so on. Strike (1982) refers to such profession-specific concepts as "intermediate-level ethical concepts," as they lie in an intermediate zone between the more general principles (e.g., autonomy, justice, and beneficence) described by philosophers and the more prescriptive directives often included in codes of conduct.

To test the possibility of designing a profession-specific test of ethical reasoning that could be used to assess the acquisition of intermediate concepts taught in a curriculum and that could be used to study the relationship between abstract reasoning and competence to reason about new professional problems, Bebeau and Thoma (1999) designed and tested the The Dental Ethical Reasoning and Judgment Test (DERJT). Similar to DIT, the test consists of five ethical problems in dentistry to which the respondent provides action choices and justification choices. The action and justification choices for each problem were generated by a group of dental faculty and residents. The scoring key reflects consensus among a national sample of 14 dental ethicists as to better, worst, and neutral choices and justification but does not prescribe a single best action or justification.

When taking the test, a respondent rates each action or justification and then selects the two best and the two worst action choices and the three best and the two worst justifications. Scores are determined by calculating the proportion of times that a respondent selects action choices and justifications consistent with "expert judgment." High levels of agreement among 14 dental ethicists as to better and worse action choices (88 percent agreement for appropriate and inappropriate actions respectively and 95 and 93 percent agreement for appropriate and inappropriate justifications, respectively) demonstrated the validity of the construct. Bebeau and Thoma (1999) reported effect sizes of 0.93 and 0.56 for action and justification choices, respectively, between first-year college students and first-year dental school students, and effect sizes of 0.85 and 0.56, respectively, between first-year dental school students and dental school seniors in the class of 1997.

Additionally, in a recent study of 308 graduates who completed DERJT and DIT, Bebeau and Thoma (2000) report that scores on DERJT are related to those on DIT ($r = 0.22$) but that the two tests are not a redundant source of information about competence in ethical reasoning and judgment. In addition, the results indicated that students with a good grasp of abstract moral schemas (good DIT P Index scores) were better able to solve the novel ethical problems presented on DERJT. As with other measures of ethical development, scores on DERJT were not related to a student's grade point average.

Identity Formation and Role Concept Development

One of the chief objectives of the study described in *On Being a Scientist* (NAS, 1989, 1995) was to convey the central values of the scientific enterprise. In an earlier era, such values were typically conveyed informally, through mentors and research advisers. Today, educators recognize the need to introduce the responsibilities more formally. Anderson (2001), in her study of doctoral students' conceptions of science and its norms, concludes that students might not be subject to as much group socialization through osmosis as many faculty assume. Nonetheless, the means by which socialization to the normative aspects of academic life are communicated are primarily informal (Anderson, 2001).

In addition to providing support for the need to more deliberately socialize students to the norms of the research enterprise, Anderson's study will likely provide grist for the design or modification of items used to assess role concept development for researchers. Such measures have been developed in some professions to assess identity formation and its relationship to ethical action.

Professional Role Orientation Inventory

The Professional Role Orientation Inventory (PROI) (Bebeau et al., 1993; Thoma et al., 1998) consists of four 10-item likert scales that assess commitment to privilege professional values over personal values. Two of the scales assess dimensions of professionalism that are theoretically linked to models of professionalism described in the professional ethics literature (e.g., Emanuel and Emanuel, 1992; May, 1983; Ozar, 1985; Veatch, 1986). The PROI scales—in particular, the responsibility and authority scales—have been shown to consistently differentiate beginning and advanced student groups and practitioner groups, who are expected to differ in their role concepts. By plotting the responses of a cohort on a two-dimensional grid (Bebeau et al., 1993), it is possible to observe four distinctly different views of professionalism that, if applied, would favor different decisions about the extent of responsibility to others.

In comparing practicing dentists with entering students and graduates, Minnesota graduates consistently express a significantly greater sense of responsibility to others than entering students and practicing dentists from the region. This finding has been replicated for five cohorts of graduates (n = 379). Additionally, the mean score for the graduates was not significantly different from that for a group of 48 dentists, who demonstrated a special commitment to professionalism by volunteering to participate in a national seminar to train individuals to be leaders of ethics seminars. A recent comparison of pretest and posttest scores for students in the classes of 1997 to 1999 (Bebeau, 2001) indicates a significant change (p < 0.0001) from the pretest to the posttest scores. Cross-sectional studies of differences between pretest and posttest scores for students in a comparable dental program suggest that instruction in ethics accounts for the change.

The most direct evidence of a relationship between role concept and professionalism comes from the study of the performances of 28 members of the practicing community referred for courses in dental ethics because of violations of the Dental Practice Act. Although the practitioners varied considerably on measures of ethical sensitivity, reasoning, and ethical implementation, 27 of the 28 individuals were unable to clearly articulate role expectations for a professional (Bebeau, 1994). (See Bebeau et al. [1993] for a more extensive description of the theoretical grounding for this measure.)

Professional Decisions and Values Test

Rezler and colleagues (1992) designed the Professional Decisions and Values Test for lawyers and physicians to assess action tendencies and the underlying values in situations with ethical problems. Patterned after DIT and the Medical Ethics Inventory, the test consists of 10 case vignettes, to which respondents provide three alternative actions and seven reasons to explain the action chosen. Actions are arranged from the least to the most intrusive, and the reasons represent one of seven values commonly used to resolve an ethical dilemma. The cases were selected to represent three themes: (1) obligation to the patient versus obligation to society, (2) respect for client autonomy versus professional responsibility, and (3) protection of the patient's interest versus respect for authority. In the presentation of the findings, data for two consecutive classes of entering medical and law students ($n = 340$) are presented, as are their action choices, and the values are compared. Although the findings support the construct validity of the test, test-retest reliability is stable over time for action choices but not for values. The developers hypothesize that values do not become stable until later in the curriculum; thus, the test may be more useful for the assessment of change over time than for the tracking of changes for individuals.

Differences by sex and profession were observed when the measure was used. Whether the lack of stability in the retest reliability study can be attributed to changes that are influenced by the curriculum is a question worthy of further study. Although further validation work needs to be done with this measure, the test is cited because its format shows promise for the design of a measure of role concept.

Ethical Implementation

In terms of the implementation of programs on professional ethics, Braxton and Baird (2001) point to the importance of providing preparation for professional self-regulation, and Fischer and Zigmond (2001) stress the importance of a variety of skills relevant to professional practice. To date, objective measures have not been devised to measure competence in the implementation of effective action plans. Although there may be some generic abilities, like problem-solving abilities and abilities in interpersonal and written communication, that could be assessed by the use of objective tests, it is hard to imagine designing anything but performance-based assessments of the broad range of skills required for effective, responsible research practice. Instructional programs could consider collecting examples of professional performance for evaluations by faculty and students, similar to the portfolios that Gilmer (1995) has students develop for her courses in research ethics. Also, institutions could draw attention to the importance of integrity in the conduct of science by including questions derived from the definition of integrity in regular faculty evaluations of research competence, including evaluations used to make promotion and tenure decisions.

SUMMARY AND CONCLUSION

A considerable amount of work has been done on the development of measurements of ethical integrity that has relevance for research institutions concerned with the assessment of integrity in the research environment. This appendix has described outcome measures and models for the development of outcome measures that address two specific purposes. The first is to assess the ethical and moral culture and climate of an institution to ensure that the climate, which includes policies and procedures related to the ethical conduct of research, supports the individual researcher's ability to function at the leading edge of professional integrity. Research in organizational behavior indicates that the ethical and moral climate of an institution can either inhibit or promote the responsible conduct of research.

The second purpose is to describe measures and methods developed in other settings of education in professional ethics that could be used directly or that could be adapted for use in the assessment of the effectiveness of courses on the responsible conduct of research or the effectiveness of an institution's efforts to promote integrity in research. The following criteria were used for the selection of measures for the latter category: the measures had to be theoretically grounded in a well-validated psychological theory of morality, were at least indirect measures of behavior, and either had been effectively used or have good potential to link the development of aspects of integrity (e.g., ethical sensitivity, moral reasoning and judgment, and identity formation) to institutional effectiveness.

In the case of methods and measures that an institution might use in a self-assessment of its moral climate, none that are directly applicable to the research setting have been developed. On the other hand, by modifying the content of the process for assessment of an institution's moral climate and the survey items used to collect information on the perceptions of individuals who work in that climate, it should be possible for an institution to gather information that would enable it to conduct an effective self-study. A reviewer of the section on the assessment of an institution's moral climate will notice that data on the psychometric properties of the surveys developed for climate assessment are not readily available for the examples described here. Given such data, it would be necessary not only to modify the content of such a survey but also to conduct appropriate validation studies.

In the case of measures for the assessment of outcomes of instruction in the responsible conduct of research, with the exception of DIT (a well-validated test of moral development over the life span that has been used effectively in intervention studies and in institutional outcome studies) the content of measures would need to be adapted. Several models for measurement have been sufficiently tested in the context of a professional ethics education program to warrant their application to the setting of integrity in research. Chapter 5 of this report gives considerable attention to teaching the responsible conduct of research. Far less attention, however, has been given to assessments of learning. One reason is the lack of well-validated outcome measures that can be used to assess the effects of instruction on the responsible conduct of research. Because individual teachers and even individual institutions are unlikely to be able to mount the kind of research and development plan needed to design and validate measures that assess the important outcomes of education in the responsible conduct of research, a national effort is needed. The design of such measures should be grounded in a well-established theory of ethical development and should be sufficiently user friendly to enable their use for a variety of purposes. Such purposes may include the following: (1) determining the range of criteria that define competence in ethical behavior in various disciplines; (2) conducting a needs assessment to identify areas where instructional resources should be placed; (3) identifying individual differences or problems that require intervention or remediation; (4) providing feedback to individuals, departments, and institutions on competence in research ethics; (5) determining the effects of current programs; (6) certifying research competence in ethical behavior; and (7) studying the relationship between competence and ethical behavior.

Given the paucity of suitable methods for the assessment of integrity in the research environment and the skepticism that education in the responsible conduct of research can make a measurable difference in important abilities related to the responsible conduct of research, there appears to be a clear need for work on the development of measurements that would serve the research community. There is also a need to design, modify, or adapt methods and survey measures to evaluate the culture and climate that promotes integrity in research.

REFERENCES

Anderson M. 2001. *What Would Get You in Trouble: Doctoral Students' Conceptions of Science and Its Norms.* Proceedings of the ORI Conference on Research on Research Integrity. [Online]. Available: http://www-personal.umich.edu/~nsteneck/rcri/index.html [Accessed March 13, 2002].

Bebeau MJ. 1994. Influencing the moral dimensions of dental practice. In: *Moral Development in the Professions: Psychology and Applied Ethics.* Hillsdale, NJ: L. Erlbaum Associates. Pp. 121–146.

Bebeau MJ. 2001. *Influencing the Moral Dimensions of Professional Practice: Implications for Teaching and Assessing for Research Integrity.* Proceedings of the ORI Conference on Research on Research Integrity. [Online]. Available: http://www-personal.umich.edu/~nsteneck/rcri/index.html [Accessed March 13, 2001].

Bebeau MJ, Brabeck MM. 1987. Integrating care and justice issues in professional moral education: A gender perspective *Journal of Moral Education* 16:189–203.

Bebeau MJ, Davis EL. 1996. Survey of ethical issues in dental research. *Journal of Dental Research* 75:845–855.

Bebeau MJ, Rest JR. 1990. *The Dental Ethical Sensitivity Test.* Minneapolis, MN: Division of Health Ecology, School of Dentistry, University of Minnesota.

Bebeau MJ, Thoma SJ. 1994. The impact of a dental ethics curriculum on moral reasoning. *Journal of Dental Education* 58:684–692.

Bebeau MJ, Thoma SJ. 1999. "Intermediate" concepts and the connection to moral education. *Educational Psychology Review* 11:343–360.

Bebeau MJ, Thoma SJ. 2000 (July 8). *The Validity and Reliability of an Intermediate Ethical Concepts Measure.* Paper presented at the annual meeting of the Association for Moral Education. Glasgow, Scotland.

Bebeau J, Rest JR, Yamoor CM. 1985. Measuring dental students' ethical sensitivity. *Journal of Dental Education* 49:225–235.

Bebeau MJ, Born DO, Ozar DT. 1993. The development of a Professional Role Orientation Inventory. *Journal of the American College of Dentists* 60(2):27–33.

Bebeau MJ, Pimple KD, Muskavitch KMT, Borden SL, Smith DL. 1995. *Moral Reasoning in Scientific Research: Cases for Teaching and Assessment.* Bloomington, IN: Indiana University. [Online]. Available: http://www.indiana.edu/~poynter/mr-main.html [Accessed March 15, 2002].

Bebeau MJ, Rest JR, Narvaez DF. 1999. Beyond the promise: A perspective for research in moral education. *Educational Researcher* 28(4):18-26.

Bowen MG, Power CP. 1993. The moral manager: Communicative ethics and the Exxon Valdez disaster. *Business Ethics Quarterly* 3:97–115.

Brabeck MM. 1998. Racial ethical sensitivity test: REST videotapes. Chestnut Hill, MA: Lynch School, Boston College.

Brabeck MM, Sirin S. 2001. *The Racial Ethical Sensitivity Test: Computer Disk Version (REST-CD).* Chestnut Hill, MA: Lynch School, Boston College.

Brabeck MM, Rogers LA, Sirin S, Henderson J, Benvenuto M, Weaver M, Ting K. 2000. Increasing ethical sensitivity to racial and gender intolerance in schools: Development of the racial ethical sensitivity test. *Ethics & Behavior* 10:119–137.

Brabeck MM, Weisgerber K. 1989. Responses to the Challenger tragedy: Subtle and significant gender differences. *Sex Roles* 19:639–650.

Braxton J, Baird L. 2001. Preparation for professional self regulation. *Science and Engineering Ethics* 7:593–614.

Burnett D, Rudolph L, Clifford K., eds. 1998. *Academic Integrity Matters.* Washington, DC: National Association of Student Personnel Administrators, Inc.

Colby A, Kohlberg L, Speicher B, Hewer A, Candee D, Gibbs J, Power C. 1987. *The Measurement of Moral Judgment*, Vols. 1 and 2. New York, NY: Cambridge University Press.

Cronbach LJ. 1951. Coefficient alpha and the internal structure of tests. *Psychometrika* 16:297–334.

Cullen J, Victor B, Stephens C. 1989. An ethical weather report: Assessing the organization's ethical climate. *Organizational Dynamics* 18:50–62.

Cullen JB, Victor B, Bronson JW. 1993. The ethical climate questionnaire: An assessment of its development and validity. *Psychological Reports* 73:667–674.

Emanuel E, Emanuel L. 1992. Four models of the physician-patient relationship. *Journal of the American Medical Association* 267:2221–2226.

Ernest M. 1990. *Developing and Testing Cases and Scoring Criteria for Assessing Geriatric Dental Ethical Sensitivity.* M.S. thesis. University of Minnesota, Minneapolis.

Fischer BA, Zigmond MJ. 2001. Promoting responsible conduct in research through "survival skills" workshops: Some mentoring is best done in a crowd. *Science and Engineering Ethics* 7:563–587.

Fleck-Henderson A. 1995. *Ethical Sensitivity: A Theoretical and Empirical Study.* Doctoral dissertation. The Fielding Institute, Santa Barbara, California.

Gibbs JC, Basinger KS, Fuller D. 1992. *Moral Maturity: Measuring the Development of Sociomoral Reflection.* Hillsdale, NJ: Erlbaum Associates.

Gilmer PJ. 1995. Teaching science at the university level: What about the ethics? *Science and Engineering Ethics* 1:173–180.

Heitman E, Salis, P, Bulger, RE 2000. Teaching ethics in biomedical sciences: Effects on moral reasoning skills. Paper presented at the ORI Research Conference on Research Integrity, Washington, D.C., November 2000 [Online]. Available http://ori.dhhs.gov/multimedia/acrobat/papers/heitman.pdf [Accessed March 15, 2002].

Higgins A, Power C, Kohlberg L. 1984. The relationship of moral atmosphere to judgments of responsibility. In: Kurtines WM, Gewirtz JL, eds., *Morality, Moral Behavior, and Moral Development*. New York, NY: Wiley. Pp. 74–108.

Howe K. 1982. Evaluating philosophy teaching: Assessing student mastery of philosophical objectives in nursing ethics. *Teaching Philosophy* 5(1):11–22.

Kohlberg L. 1984. *The Psychology of Moral Development: The Nature and Validity of Moral Stages*. Essays on Moral Development Vol. 2. San Francisco: Harper & Row.

Korenman SG, Berk R, Wenger NS, Lew V. 1998. Evaluation of the research norms of scientists and administrators responsible for academic research integrity. *Journal of the American Medical Association* 279:41–47.

Leibowitz S. 1990. *Measuring Change in Sensitivity to Ethical Issues in Computer Use*. Doctoral dissertation. Boston College, Boston, MA.

Lind R. 1997. Ethical sensitivity in viewer evaluations of a TV news investigative report. *Human Communication Research* 23:535–561.

Lind G, Wakenhut R. 1985. Testing for moral judgment competence. In: Lind G, Hartmann HA, Wakenhut R, eds. *Moral Development and the Social Environment*. Chicago, IL: Precedent. Pp. 79–105.

May WE. 1983. *The Physician's Covenant: Images of the Healer in Medical Ethics*. Philadelphia: Westminster Press.

McAlpine H, Kristjanson L, Poroch D. 1997. Development and testing of the ethical reasoning tool (ERT): An instrument to measure the ethical reasoning of nurses. *Journal of Advanced Nursing* 25:1151–1161.

McNeel SP 1990. Development of a measure of moral sensitivity for college students. In: *Teaching Values Across the Curriculum*. Project report. Dunbarton, NH: The Christian College Consortium.

Mentkowski M. 2000. *Learning That Lasts: Integrating Learning, Development, and Performance in College and Beyond*. San Francisco, CA: Jossey-Bass.

Mentkowski M, Loacker G. 1985. Assessing and validating the outcomes of college. In Ewell PT, ed. *Assessing Educational Outcomes. New Directions for Institutional Research*. No. 47. San Francisco: Jossey-Bass. Pp. 47–64.

NAS (National Academy of Sciences). 1989. *On Being a Scientist*. Washington, DC: National Academy Press.

NAS. 1995. *On Being a Scientist*, 2nd ed. Washington, DC: National Academy Press.

OGE (U.S. Office of Government Ethics). 2000. Executive Branch Employee Ethics Survey 2000. [Online]. Available http://www.usoge.gov/pages/forms_pubs_otherdocs/fpo_files/surveys_ques/srvyemp_if_00.pdf [Accessed March 15, 2002].

Ozar DT. 1985. Three models of professionalism and professional obligation in dentistry. *Journal of the American Dental Association* 110:173–177.

Pascarella ET, Terenzini PT. 1991. Moral development. In: *How College Affects Students: Findings and Insights from Twenty Years of Research*, San Francisco, CA: Jossey-Bass. Pp. 335–368.

Ponemon, LA, Gabhart, DRL. 1994. Ethical reasoning research in the accounting and auditing professions. Rest JR, Narvaez D, ed. *Moral development in the professions: Psychology and applied ethics*. Hillsdale, NJ: Lawrence Erlbaum Associates, Inc. Pp. 101–119.

Power C. 1980. Evaluating just communities: Toward a method of assessing the moral atmosphere of the school. In: Moser R, ed. *Moral Education: A First Generation of Research and Development*. New York, NY: Praeger. Pp. 223–265.

Power C. Higgins A Kohlberg L. 1989. *Lawrence Kohlberg's Approach to Moral Education*. New York, NY: Columbia University Press.

Rest J. 1983. Morality. In: Mussen PH (series ed.) and Flavell J, Markman E (vol. eds.). *Handbook of Child Psychology*, Vol. 3, *Cognitive Development*, 4th ed. New York, NY: Wiley. Pp. 556–629.

Rest J, Narváez D, Bebeau MJ, Thoma SJ. 1999a. *Postconventional Moral Thinking: A Neo-Kohlbergian Approach*. Hillsdale, NJ: L. Erlbaum Associates.

Rest J, Narváez D, Thoma SJ, Bebeau MJ. 1999b. DIT2: Devising and testing a revised instrument of moral judgment. *Journal of Educational Psychology* 91(4):644–659.

Rest JR. 1979. *Development in Judging Moral Issues*. Minneapolis: University of Minnesota Press.

Rest JR, Narváez DF, eds. 1994. *Moral Development in the Professions: Psychology and Applied Ethics*. Hillsdale, NJ: Erlbaum Associates. Pp. 51–70.

Rest J, Thoma SJ, Narváez D, Bebeau MJ. 1997. Alchemy and beyond: Indexing the Defining Issues Test. *Journal of Educational Psychology* 89(3):498–507.

Rest JR, Bebeau MJ, Volker J. 1986. An overview of the psychology of morality. In: Rest JR, eds. *Moral Development: Advances in Research and Theory*, Boston, MA: Prager Publishers. Pp. 1-39.

Rezler AG, Schwartz RL, Obenshain SS, Lambert P, McGibson J, Bennahum DA. 1992. Assessment of ethical decisions and values. *Medical Education* 26:7–16.

Sirin S, Brabeck MM, Satiani A, Rogers LA. Submitted for publication. Development of computerized racial ethical sensitivity test.

Stern J, Elliott D. 1997. *The Ethics of Scientific Research: A Guidebook for Course Development*. Hanover, NH: University Press of New England.

Strike KA. 1982. *Educational Policy and the Just Society*. Chicago, IL: University of Chicago.

Thoma SJ, Bebeau MJ, Born DO. 1998. Further analysis of the Professional Role Orientation Inventory. *Journal of Dental Research* 77(Special Issue):120 (abstract 116).

U.S. Army. 2001. *Ethical Climate Assessment Survey*. Document GTA 22-6-1. [Online]. Available: http://www.leadership.army.mil/leaderphilosophyandvision/ECAS.htm [Accessed June 20, 2001].

Veatch RM. 1986. Models for ethical medicine in a revolutionary age. In: Mappes TA, Zembaty J, eds. *Biomedical Ethics*, 2nd ed. New York:McGraw-Hill.

Victor B., Cullen JB. 1988. The organizational bases of ethical work climates. *Administrative Science Quarterly* 33:101–125.

Volker JM. 1984. *Counseling Experience, Moral Judgement, Awareness of Consequences, and Moral Sensitivity in Counseling Practice*. Doctoral thesis. University of Minnesota, Minneapolis, MN.

Yeap CH. 1999. *An Analysis of the Effects of Moral Education Interventions on the Development of Moral Cognition*. Doctoral dissertation. University of Minnesota, Minneapolis, MN.

C

Developments in Misconduct and Integrity Policies Since Publication of the 1992 COSEPUP Report[1]

In 1992, the Committee on Science, Engineering, and Public Policy (COSEPUP) of the National Academy complex[2] published *Responsible Science: Ensuring the Integrity of the Research Process* (NAS, 1992). Its publication followed years of broader political turmoil over integrity in research and misconduct. In the 10 years since publication of the COSEPUP report there have been significant changes in the science policy system that oversees and promotes integrity in research, particularly in the Office of Research Integrity (ORI) of the U.S. Department of Health and Human Services (DHHS). The former "crisis" of integrity in research has been normalized over the last decade. A clear indicator of this normalization is the emergence of educational efforts in the responsible conduct of research and the study of integrity in research as a field unto itself. The reorganization of ORI as a less controversial body principally interested in education, oversight, assurance, and, now, research was important in this process. The transformation of the policy environment has, however, been slow.

---

[1]This chapter is based on a commissioned review prepared by David H. Guston, associate professor and director, Program in Public Policy, E. J. Bloustein School of Planning and Public Policy, Rutgers, the State University of New Jersey.

[2]The National Academy complex consists of the National Academy of Sciences, the National Academy of Engineering, and the Institute of Medicine.

STATUS QUO IN 1991

In 1991 there was still a great deal of turmoil in policies regarding misconduct in research as the COSEPUP Panel on Scientific Responsibility and the Conduct of Research completed its report. The Office of Scientific Integrity (OSI), whose creation in 1989 by then National Institutes of Health (NIH) Director James Wyngaarden was instigated by Representative John Dingell's inquiries, was assaulted by charges of incompetence and illegitimacy. OSI suffered a "scientific backlash" (Hamilton, 1991, p. 1084) that criticized it as too "zealous" (Davis, 1991, p. 12) and staffed by investigators reminiscent of the "Keystone Cops" (Wheeler, 1991, p. A5).

While testifying before Dingell's Oversight Subcommittee the subsequent NIH director, Bernadine Healy, expressed her doubts about "due process, confidentiality, fairness and objectivity" at OSI. Dingell countered that Healy had "made a mockery of the OSI's alleged independence in dealing with misconduct allegations" (Greenberg, 1991, p. 5).

A suit filed by James Abbs, a neurophysiologist at the University of Wisconsin and the subject of an investigation by OSI, asked the court to halt the investigation, charging that OSI failed to provide due process and to promulgate its procedures under the requirements of the Administrative Procedures Act (APA). Abbs argued that he had a property stake in his grant, his academic position, and his reputation and that OSI had deprived him of this property without due process of law under the Fifth and Fourteenth Amendments to the U.S. Constitution. The government argued that no such property rights exist and that, even if they did, OSI provided due process. The government also argued that OSI was not required to fulfill any requirements of public notice for its internal procedures. The district judge decided in Abbs's favor, invalidating OSI's internal policies and procedures in the Western District of Wisconsin. The judge decided the due process claim, however, in favor of the government, declaring that the invalid procedures did, in fact, provide sufficient due process. Both parties appealed the split decision, and the Seventh Circuit Court vacated the district court's ruling, validating OSI's claim under APA as well as its due process claim. Abbs and ORI ultimately accepted a settlement imposing special conditions on federal research funding. The case focused a great deal of scrutiny on OSI; however, many commentators misinterpreted the district court's opinion as substantively critical of OSI (Guston, 2000).

ORI AND THE DEPARTMENTAL APPEALS BOARD

Despite the validation of OSI by the circuit court, on February 1992, then Assistant Secretary of Health James Mason forwarded to then Secretary of Health and Human Services (HHS) Louis Sullivan a plan to reorganize DHHS's approach to scientific integrity. Under the new plan, ORI would replace both OSI and the Office of Scientific Integrity Review (OSIR). The new structure demonstrated greater attention to traditional legal concerns, as a branch of the Office of the General Counsel was incorporated in ORI.

Still responding to criticism of ORI's procedures, the U.S. Public Health Service (PHS) in November 1992 issued notice of an interim procedure under which individuals found to have committed misconduct could request an administrative hearing before the Research Integrity Adjudications Panel (RIAP) of the Departmental Appeals Board (DAB) (PHS, 1992). During such hearings, respondents could be represented by counsel, question evidence and cross-examine witnesses, and present rebuttal evidence and witnesses. Despite these enhanced procedural aspects, however, no formal rules of evidence apply to DAB hearings. The first such hearing occurred in June 1993.

Generally, DAB has been a defender of PHS action on misconduct. In response to appeals made before it, DAB has confirmed that DHHS has had the authority to investigate allegations of misconduct involving federal funds and take administrative action when misconduct has been found. DAB identifies this authority as emanating from the nature of the grants process that, as the court in the Abbs case held, is discretionary to the secretary of DHHS. DAB concluded that previous attempts to deal with misconduct, including the 1980 rules for debarment and the 1986 policies and procedures, were appropriate expressions of this authority. Under the same logic, DAB also ruled that DHHS might place conditions on the future awarding of grants and other aspects of a researcher's involvement with DHHS programs (OASH, 1994, p. 6).

One difficulty, however, has been DAB's interpretation of the burden and standard of proof, standards of conduct, and intent in research misconduct cases. DAB's ruling on the burden and standard of proof required for a finding of misconduct is straightforward: the burden rests on ORI to demonstrate misconduct by a preponderance of the evidence, which is the normal standard in civil cases but lower than the standard in the 1980 debarment rule (see 45 CFR Part 76). More recently, the DHHS Review Group on Research Misconduct endorsed the preponderance standard as well (ORI, 1999).

With respect to standards of conduct, however, DAB has held that ORI must demonstrate that the respondent's actions violated standards in effect at the time of the conduct—standards derived either from the relevant scientific community or from federal requirements of conduct. Finally, ORI must also demonstrate that the violation of standards was intentional, that is, that any reasonable researcher in the respondent's position would have understood the actions as constituting misconduct (OASH, 1994; ORI, 1993).

The introduction of the intent standard by DAB led ORI to drop charges against Robert Gallo in November 1993. DAB also overturned findings of misconduct in high-profile cases against Ramesh Sharma (in 1993), Mikulas Popovic (in 1993), and Thereza Imanishi-Kari (in 1996). The original findings of misconduct in these cases were made under OSI's policies and procedures. The ORI case record of scientific misconduct from its origin in June 1992 through the end of CY 2001 shows ORI made 125 findings of scientific misconduct. Since 1996, one case went to a full DAB hearing, which was won by ORI, leading to an HHS debarment of Dr. Kimon Angelides in 1999 from federal funding for five years (see http://ori.hss.gov/html/misconduct/ori_summary_angelides.asp). A second was settled by the respondent, in the middle of the DAB hearing; Dr. Evan Dreyer was debarred in 2000 by HHS for 10 years (see http://ori.dhhs.gov/html/programs/fedregnotice.asp).

In June 1994, the Office of the Secretary of DHHS issued notice that ORI had revised its guidelines for such hearings (DHHS, 1994). One of the revisions allowed for a scientist to be included at the request of either DHHS or the respondent (in the original procedure, a scientist would be included at the discretion of the panel chair), and in 1999 the DHHS Review Group on Research Misconduct and Research Integrity recommended that up to two scientists be allowed to serve on DAB panels (ORI, 1999).

COMMISSION ON RESEARCH INTEGRITY

In the NIH Revitalization Act of 1993 (PL 103-43), the U.S. Congress authorized ORI in law and delegated the formulation of a definition of misconduct, among other tasks, to a department-level commission of 12 members, including academic physicians, biomedical researchers, lawyers, and ethicists. The commission solicited input from the research community and reported its findings in November 1995 (DHHS, 1995). It placed an emphasis on whistle-blowing and promoted "responsible whistle-blowing" and a "whistle-blower's bill of rights" (DHHS, 1995, pp. 21–24). Implementing the commission's perspective, ORI developed *Guidelines for Institutions and Whistleblowers: Responding to Possible Retaliation Against Whistleblowers in Extramural Research* in November 1995. ORI issued a notice of proposed rulemaking in 2000 to apply due process protections to whistle-blowers at universities and is reviewing comments received (DHHS, 2000a).

The commission also addressed the definition of research misconduct, which had been controversial despite PHS's publication in 1989 of the final rule, which defined it as:

> [f]abrication, falsification, plagiarism, or other practices that seriously deviate from those that are commonly accepted within the scientific community for proposing, conducting, or reporting research. It does not include honest error or honest differences in interpretations or judgments of data (DHHS, 1995, p. 1).

At issue was the phrase "other practices. . . " and the lack of further definition of fabrication, falsification, or plagiarism. After significant deliberation, the commission recommended the following language:

> Research misconduct is significant misbehavior that improperly appropriates the intellectual property or contribution of others, that intentionally impedes the progress of research, or that risks corrupting the scientific record or compromising the integrity of scientific practices. Such behaviors are unethical and unacceptable in proposing, conducting, or reporting research, or in reviewing the proposals or research reports of others (DHHS, 1995, p. 13).

The commission further defined misappropriation, interference, and misrepresentation. It recommended that a federal interagency task force consider drafting a definition that would be common across all federal research.

THE COMMON DEFINITION[3]

In April 1996 the move toward a common definition began. The need was threefold: some research-funding agencies still lacked definitions and policies; the definitions at PHS and the National Science Foundation were not the same, leading to the possibility that a researcher jointly funded by the two agencies could be judged by different standards; and courts could conceivably overturn rulings on due process grounds, as the absence of a clear federal statement could be seen as lack of appropriate notice. The National Science and Technology Council (NSTC), an operating arm of the White House Office of Science and Technology Policy responsible for coordinating policy among the government's many agencies that perform research and development, created a panel to draft the definition.

The panel reported its definition in December 1996, followed by a round of review and comments by the various federal agencies and a public comment period before publishing it in December 2000 (Francis, 1999; OSTP, 1999). The final policy included the formation of an NSTC

---

[3]Most of this section is drawn from Guston (1999); but see Francis (1999), OSTP (1999), and Porter and Dustira (1993).

research misconduct policy implementation group; clarification or elaboration of some specific wording in the new definition; and greater specification of options and duties toward the scientific record, human research subjects, and others in response to findings of misconduct (OSTP, 2000). The final common policy defined research misconduct as

> fabrication, falsification, or plagiarism in proposing, performing, or reviewing research, or in reporting research results. . . . Fabrication is making up of data or results and recording or reporting them. Falsification is manipulating research materials, equipment, or processes, or changing or omitting data or results such that the research is not accurately represented in the research record. . . . Plagiarism is the appropriation of another person's ideas, processes, results, or words without giving appropriate credit. Research misconduct does not include honest error or differences of opinion (OSTP, 2000, p. 76262). A finding of research misconduct requires that: There be a significant departure from accepted practices of the relevant research community; and the misconduct be committed intentionally, or knowingly, or recklessly; and the allegation be proven by a preponderance of the evidence (65 Fed. Reg. 76260-76264).

Although the definition and other content of the rule are final, they are not effective until agencies formally implement the rule, a process expected to be complete in 2003.

In the meantime, the Wellcome Trust, the largest biomedical charity in the United Kingdom, offered its own definition of scientific misconduct:

> [t]he fabrication, falsification, plagiarism or deception in proposing, carrying out or reporting results of research or deliberate, dangerous or negligent deviations from accepted practices in carrying out research. It includes failure to follow established protocols if this failure results in unreasonable risk or harm to humans, other vertebrates or the environment (Koenig, 2001, p. 1411).

This definition, with its inclusion of "negligent deviations" and attention to risks to research subjects, is broader than the U.S. definition. In light of this development, some observers have noted the possible benefits of international discussions and even international guidelines for research ethics (Kaiser, 1999; Mishkin, 1999).

RESPONSIBLE CONDUCT OF RESEARCH

In 1996, DHHS Secretary Donna Shalala charged Assistant Secretary of Health Philip Lee to lead a group of DHHS officials in reviewing the department's misconduct procedures and the commission's recommendations. This group's review ended ORI's role in investigations of research misconduct (DHHS, 1999). Grantee institutions would continue to conduct investigations, and should federal investigations be needed, the Office of the Inspector General would conduct them (DHHS, 2000b). ORI would also no longer be responsible for conducting investigations for PHS intramural laboratories. ORI's primary function would be to support grantee institutions through educating officials from such institutions, validating their policies for investigating misconduct, overseeing their findings in specific cases, and generally bolstering their legitimacy to maintain some degree of control over integrity in research (ORI, 2000a).

The responsible conduct of research had to some extent already been a focus of important developments. As early as 1990, NIH began by requiring a plan for education in research ethics as part of the application for NIH research training grants. NIH did not establish particular curricula or curricular requirements, and the quality of the curriculum did not contribute to the score for the grant application, but applications could not be funded until acceptable plans were articulated (Dustira, 1996).

More recently, ORI proposed a policy to require all funded institutions to provide educational programs in the responsible conduct of research for all research staff associated with PHS funds. Several groups representing institutions of higher education issued a "community comment" in response to the proposal (Hasselmo et al., 2000). The comment "state[d] unequivocally" the support of these groups for the "core value" of "integrity of research and teaching" and a "strong belief in promoting the responsible conduct of research and preventing research misconduct through education and awareness" (Hasselmo et al., 2000, p.1). Nevertheless, the group thought that the proposal had an "overly prescriptive tone...which resembles rulemaking more than policy" (p.2). It took issue with the "heavy-handed prescription" (p.3) of the core instructional areas and made specific recommendations to allow institutions to have greater flexibility in specifying both the curriculum and the definition of research staff to whom the policy would apply.

According to ORI, the final policy took these concerns into consideration by providing institutions with "considerable flexibility in designing an educational program for their research staff and extends the implementation period to October 1, 2003" (ORI, 2000b, p. 1). In February 2001, however, DHHS suspended the implementation of these requirements at the behest of the House Energy and Commerce Committee, which suggested that the department did not adhere to APA requirements in promulgating them. The Energy and Commerce Committee did not, however, explicitly question the substance of the rules (Brainard, 2001).

Although there have been meetings among interested parties about

---

BOX C-1
Time Line of Some Significant Events in Research Integrity,
1991 to Present

1991

COSEPUP Panel on Scientific Responsibility and the Conduct of Research holds final meeting.
PHS Advisory Committee on Scientific Integrity meets for the first time.
Congressional hearings held on investigation of research misconduct.

1992

Office of Research Integrity (ORI) is created within the Office of the Assistant Secretary for health by merging the Office of Scientific Integrity (OSI) and Office of Scientific Integrity Review (OSIR).
PHS announces interim procedures for hearings before the Departmental Appeals Board (DAB).
NIH strengthens responsible conduct of research requirement in training grant applications.

1993

PHS Advisory Committee on Scientific Integrity holds last meeting.
DAB holds first hearing.
PHS ALERT system listings limited to misconduct findings; allegations no longer included.
DAB confirms HHS authority to investigate scientific misconduct.
NIH Revitalization Act codifies the establishment of ORI, creates the Commission on Research Integrity, and mandates the development of a regulation to protect whistle-blowers.
ORI begins publishing information about closed cases of confirmed research misconduct.
Commission on Research Integrity chartered.
ORI drops pursuit of allegations against Robert Gallo.

1994

ORI revises guidelines for inclusion of scientists on DAB and other procedures for DAB hearings.
Notification to journal editors about corrections or retractions resulting from confirmed research misconduct is initiated by ORI.
Reviews of allegations of retaliation against whistle-blowers start.
Model Policy and Procedures are developed for responding to research misconduct allegations.

1995

Commission on Research Integrity, under Kenneth Ryan, issues report.
ORI publishes Guidelines for Institutions and Whistleblowers: Responding to Possible Retaliation Against Whistleblowers in Extramural Research.
ORI reviews of institutional policies for responding to research misconduct allegations begin.

1996

Departmental reorganization places ORI in the Office of Public Health and Science.

Office of Science and Technology Policy (OSTP) begins to develop common federal definition of research misconduct.

DAB reverses misconduct finding in Imanishi-Kari case.

DHHS Review Group on Research Misconduct and Research Integrity is created.

1998

ORI reports that 174 institutions reported 432 allegations of research misconduct from 1991-1996, and that investigations by ORI from 1993-1997 resulted in 76 misconduct findings and 74 no-misconduct findings.

1999

DAB upholds research misconduct finding in Angelides case.

HHS Secretary Shalala implements more than a dozen recommendations from the HHS Review Group on Research Misconduct and Research Integrity.

Investigation of research misconduct allegations transferred from ORI to institutions, PHS agencies, Office of the Inspector General and research institutions.

ASH delegates responsibility for research misconduct in intramural laboratories to the heads of PHS agencies, with ORI providing oversight.

ORI mission is refocused on oversight, education, and prevention.

2000

First ORI research conference on research integrity is held.

Guidance for editors managing research misconduct allegations is published by ORI.

Research program on research integrity is initiated by ORI in collaboration with National Institute of Neurological Disorders and Stroke.

PHS Policy on Instruction in the Responsible Conduct of Research is published.

Federal Research Misconduct Policy, including common definition of misconduct, is published by OSTP.

Notice of proposed rule making on protection of whistleblowers published by HHS.

2001

First awards made in research on research integrity program.

PHS Policy on Instruction in the Responsible Conduct of Research suspended.

Institute of Medicine Committee on Assessing Integrity in Research Environments holds initial meeting.

the requirements for the responsible conduct of research, a final resolution has not occurred. In the interim, ORI has also established a listserv on which individuals wishing to discuss the responsible conduct of research can share information and experience.

The decade since publication of the COSEPUP report has witnessed a great deal of activity in the development and evolution of policies regarding research misconduct. Since the early 1990s the federal government has asserted its authority and discretion in setting conditions on the awarding of research grants. It requires research institutions to have policies and procedures in place for handling allegations of misconduct, protecting whistle-blowers, and providing training in research ethics in training grants. At the same time that research institutions have augmented their ability to combat misconduct, the specific role of the government in investigating allegations has been legalized and rarefied. The government, particularly through the DHHS DAB, has articulated clear standards for the adjudication of allegations. ORI has proposed to DHHS a revised regulation, consistent with the 2000 OSTP policy, to replace the 1989 PHS rule. Thus, the new, government-wide definition of research misconduct and associated procedures should be implemented shortly. The definition is more precise than previous definitions and is both narrower and more expansive in different areas. Its precision, however, seems to buttress the greater emphasis on due process, particularly the emphasis on notice, that has been evolving as well. ORI is pursuing its charge to prevent misconduct and promote research integrity by maintaining oversight over institutional research misconduct investigations, providing technical assistance to institutions handling allegations, defending research misconduct findings before the DAB, facilitating the creation of RCR programs at institutions, developing a research program on research integrity, responding to retaliation complaints from whiste-blowers, and ensuring regulatory compliance.

REFERENCES

Brainard J. 2001, March 2. Ethics-training rule is suspended. *The Chronicle of Higher Education.* P. A27.
Davis B. 1991, May 13. Is the Office of Scientific Integrity too zealous? *The Scientist.* 5(10):12.
DHHS (U.S. Department of Health and Human Services). 1994. Hearing procedures for scientific misconduct. *Federal Register* 59:29808–29811.
DHHS. 1995. *Integrity and Misconduct in Research: Report of the Commission on Research Integrity.* Rockville, MD: Office of Research Integrity, Office of the Secretary, DHHS.
DHHS. 1999. *Report of the Department of Health and Human Services Review Group on Research Misconduct and Research Integrity.* [Online]. Available: http://ori.hhs.gov/html/publications/dhhsreview2.asp#background [Accessed March 14, 2002].
DHHS. 2000a. Notice of proposed rulemaking: Public Health Service standards for the protection of research misconduct whistleblowers. *Federal Register* 64:70830–70841.

DHHS. 2000b. Statement of organization, functions, and delegations of authority. *Federal Register* 65:30600–30601.

Dustira AK. 1996. The federal role in influencing research ethics education and standards in science. *Professional Ethics* 5(1-2):139–156.

Francis S. 1999. Developing a federal policy on research misconduct. *Science and Engineering Ethics* 5:261–272.

Greenberg DS. 1991. Q&A with NIH Director Bernadine Healy. *Science & Government Report* 5(1):5.

Guston DH. 1999. Changing explanatory frameworks in the U.S. government's attempt to define research misconduct. *Science and Engineering Ethics* 5:137–154.

Guston DH. 2000. *Between Politics and Science: Assuring the Integrity and Productivity of Research.* New York, NY: Cambridge University Press.

Hamilton DP. 1991. Can OSI withstand a scientific backlash? *Science* 253:1084–1086.

Hasselmo N, Stewart D, Phillips K, Magrath CP. 2000. *Community Comment on PHS Research Instruction Policy.* Letter to Chris B. Pascal, ORI, September 21. [Online]. Available: http://www.aau.edu/research/phs9.21.00.html [Accessed March 14, 2002].

Kaiser M. 1999. Development of international guidelines for research ethics. *Science and Engineering Ethics* 5:293–298.

Koenig R. 2001. Wellcome rules widen the net. *Science* 293:1411–1413.

Mishkin B. 1999. Scientific misconduct: Present problems and future trends. *Science and Engineering Ethics* 5:283–292.

NAS (National Academy of Sciences). 1992. *Responsible Science: Ensuring the Integrity of the Research Process,* Vol. I. Committee on Science, Engineering, and Public Policy, Panel on Scientific Responsibility and the Conduct of Research. Washington, DC: National Academy Press.

OASH (Office of the Assistant Secretary of Health). 1994. *Office of Research Integrity: Annual Report, 1993.* Washington, DC: OASH.

ORI (Office of Research Integrity). 1993. DAB confirms HHS authority to investigate scientific misconduct. *ORI Newsletter* 1(4):5.

ORI. 1997. *Annual Report, 1996.* U.S. Department of Health and Human Services, Bethesda, MD. [Online]. Available: http://ori.dhhs.gov/html/publications/annual-reports.asp#1996 [Accessed March 14, 2002].

ORI. 1999. Review group recommendations being implemented rapidly. *ORI Newsletter* 8(1):5–6.

ORI. 2000a. ORI director chosen and other senior appointments made. *ORI Newsletter* 8(4):7.

ORI. 2000b. Final RCR policy provides flexibility and more time to institutions. *ORI Newsletter* 9(1):1–2.

OSTP (Office of Science and Technology Policy). 1999. Proposed federal policy on research misconduct to protect the integrity of the research record. *Federal Register* 64:55722–55725.

OSTP. 2000. Federal policy on research misconduct. *Federal Register* 65:76260–76264.

Porter JP, Dustira AK. 1993. Policy development lessons from two federal initiatives: Protecting human research subjects and handling misconduct in science. *Academic Medicine* 68:S51–S55.

PHS (Public Health Service). 1992. Opportunity for a hearing on Office of Research Integrity scientific misconduct findings. *Federal Register* 57:53125–53126.

Wheeler DL. 1991, May 15. NIH office that investigates scientists' misconduct is target of widespread charges of incompetence. *The Chronicle of Higher Education.* P. A5.

D

Additional Resources Regarding Professional Skills

A number of books and articles have been published on aspects of survival skills, ethics, and the responsible conduct of research. The following list is provided as a starting point for readers seeking additional resources. The committee does not endorse any particular programs or recommendations from any of the publications listed below. Additional resources may also be found online at www.edc.gsph.pitt.edu/survival/resources2.html.

Being an Apprentice, Doing Science, Creativity

Beveridge WIB. 1950. *The Art of Scientific Investigation*. London: Vintage Books.

Gardner H. 1993. *Creating Minds: An Anatomy of Creativity Seen Through the Lives of Freud, Einstein, Picasso, Stravinsky, Eliot, Graham and Gandhi*. New York, NY: Basic Books.

Grinnell F. 1992. *The Scientific Attitude*, 2nd ed. New York, NY: Guilford Press.

Medawar PB. 1979. *Advice to a Young Scientist*. New York, NY: Basic Books.

National Academy of Sciences, National Academy of Engineering, and Institute of Medicine. 1996. *Careers in Science and Engineering: A Student Planning Guide to Grad School and Beyond*. Committee on Science, Engineering, and Public Policy. Washington, DC: National Academy Press.

National Academy of Sciences, National Academy of Engineering, and Institute of Medicine. 2000. *Enhancing the Postdoctoral Experience for Scientists and Engineers: A Guide for Postdoctoral Scholars, Advisers, Institutions, Funding Organizations, and Disciplinary Societies.* Committee on Science, Engineering, and Public Policy. Washington, DC: National Academy Press.

Peters RL. 1997. *Getting What You Came for: The Smart Student's Guide to Earning a Master's or a Ph.D.* New York, NY: Farrar, Straus, and Giroux.

Rogoff B. 1990. *Apprenticeship in Thinking.* New York, NY: Oxford University Press.

Shekerijian D. 1991. *Uncommon Genius: How Great Ideas Are Born.* New York, NY: Penguin Books.

Grantspersonship

Miner LE, Griffith J. 1993. *Proposal Planning & Writing.* Phoenix, AZ: Oryx Press.

Reif-Lehrer L. 1995. *Grant Application Writer's Handbook.* Boston, MA: Jones and Bartlett Publishers.

Ries JB. Leukefeld CG. 1995. *Applying for Research Funding: Getting Started and Getting Funded.* Thousand Oaks, CA: Sage Publications.

Graphics

Briscoe MH. 1996. *Preparing Scientific Illustrations: A Guide to Better Posters, Presentations, and Publications.* New York, NY: Springer-Verlag.

Job Hunting

Bolles RN. 2001 (updated yearly). *What Color Is Your Parachute? A Practical Manual for Job Hunters and Career Changers.* Berkeley, CA: Ten Speed Press.

Fisher R, Ury W. 1991. *Getting to Yes,* 2nd ed. New York, NY: Penguin Books.

National Academy of Sciences. 1996. *Careers in Science and Engineering: A Student Planning Guide to Grad School and Beyond.* Washington, DC: National Academy Press. (Available for free download at the NAS website: http://www.nationalacademies.org/publications/.)

Yate M. 2002. *Knock 'em Dead 2002: The Ultimate Job Seeker's Resource.* Holbrook, MA: Adams Media.

Mentoring

Fort C, Bird SJ, Didion CJ, eds. 1993. *A Hand Up: Women Mentoring Women in Science.* Washington, DC: Association for Women in Science.

Freire P, with Fraser JW, Macedo D, McKinnon T, Stokes WT, eds. 1997. *Mentoring the Mentor: A Critical Dialogue with Paulo Freire.* New York, NY: Peter Lang Publishing, Inc.

Kanigel R. 1986. *Apprentice to Genius: The Making of a Scientific Dynasty.* Baltimore, MD: Johns Hopkins University Press.

National Academy of Sciences, National Academy of Engineering, and Institute of Medicine. 1997. *Adviser, Teacher, Role Model, Friend: On Being a Mentor to Students in Science and Engineering.* Washington, DC: National Academy Press.

Oral Communications

Schloff L., Yudkin M. 1992. *Smart Speaking.* New York, NY: Plume.

Stuart C. 1989. *How to Be an Effective Speaker.* Chicago, IL: NTC Publishing Group.

Personal and Professional Development

Griessman BE. 1994. *Time Tactics of Very Successful People.* New York, NY: McGraw-Hill.

Hobfoll SE, Hobfoll IH. 1994. *Work Won't Love You Back: The Dual Career Couple's Survival Guide.* New York, NY: W.H. Freeman.

Roesch R. 1996. *The Working Woman's Guide to Managing Time.* Englewood Cliffs, NJ: Prentice-Hall.

Providing Access

Barba RH, Reynolds KE. 1998. Towards an equitable learning environment in science for Hispanic students. In: Fraser B, Tobin KG, eds. *International Handbook of Science Education*, Part 2. Dordrecht, The Netherlands: Kluwer Academic Publishers. Pp. 925–939.

Belenky MF, Clinchy BM, Goldberger NR, Tarule JM. 1986 (reissued in 1997). *Women's Ways of Knowing: The Development of Self, Voice and Mind.* New York, NY: Basic Books.

Gallard A, Viggiano E, Graham S, Stewart G, Vigliano M. 1998. The learning of voluntary and involuntary minorities in science classrooms. In: Fraser B, Tobin KG, eds. *International Handbook of*

Science Education, Part 2. Dordrecht, The Netherlands: Kluwer Academic Publishers. Pp. 941–953.

Hewson PW, Beeth ME, Thorley NR. 1998. Teaching for conceptual change. In: Fraser B,Tobin KG, eds. *International Handbook of Science Education*, Part 1. Dordrecht, The Netherlands: Kluwer Academic Publishers. Pp. 199–218.

Katz M, Vieland V. 1993. *Get Smart! What You Should Know (but Won't Learn in Class) About Sexual Harassment and Sexual Discrimination.* New York, NY: The Feminist Press, City University of New York.

Kucera TJ, ed. 1993. *Teaching Chemistry to Students with Disabilities*, 3rd ed. Washington, DC: American Chemical Society.

Ladson-Billings G. 1995. Toward a theory of culturally relevant pedagogy. *American Educational Research Journal* 32:465–491.

Matyas ML, Malcom SM, eds. 1991. *Investing in Human Potential: Science and Engineering at the Crossroads.* Washington, DC: American Association for the Advancement of Science.

Mitchell R. 1993. *The Multicultural Student's Guide to Colleges.* New York, NY: Noonday Press.

National Science Foundation. 1997. *Women and Science: Celebrating Achievements, Charting Challenges.* Arlington, VA: National Science Foundation.

Nichols S, Gilmer PJ, Thompson T, Davis N. 1998. Women in science: expanding the vision. In: Fraser B, Tobin KG, eds. *International Handbook of Science Education*, Part 2. Dordrecht, The Netherlands: Kluwer Academic Publishers. Pp. 967–978.

Parker LH, Rennie LJ, Fraser BJ, eds. 1996. *Gender, Science and Mathematics: Shortening the Shadow.* Dordrecht, The Netherlands: Kluwer Academic Publishers.

Posner GJ, Strike KA, Hewson PW, Gertzog WA. 1982. Accommodation of a scientific conception toward a theory of conceptual change. *Science Education* 66:211–227.

Shepherd LJ. 1993. *Lifting the Veil. The Feminine Face of Science.* Boston, MA: Shambhala.

Sonnert G, Holton G. 1995. *Who Succeeds in Science? The Gender Dimension.* New Brunswick, NJ: Rutgers University Press.

Responsible Scientific Conduct

Bebeau MJ, Pimple KD, Muskavitch KMT, Borden SL, Smith DL. 1995. *Moral Reasoning in Scientific Research: Cases for Teaching and Assessment.* Bloomington, IN: Indiana University.

Bulger RE, Heitman E, Reiser SJ, eds. 1993. *The Ethical Dimensions of the Biological Sciences.* New York: Cambridge University Press.

Edsall JT. 1981. Two aspects of scientific responsibility. *Science* (4490):11-14. (Reprinted in: Chalk R, ed. 1988. *Science, Technology, and Society: Emerging Relationships.* Washington, DC: American Association for the Advancement of Science. Pp. 12–15.)

Glass B. 1965. The ethical basis of science. *Science* (150)3701:1254-1261. (Reprinted in: Chalk R, ed. 1988. *Science, Technology, and Society: Emerging Relationships.* Washington, DC: American Association for the Advancement of Science. Pp. 37–43.)

Grinnell F. 1997. Truth, fairness, and the definition of scientific misconduct. *Journal of Laboratory and Clinical Medicine* 129:189–192.

Grinnell F. 1999. Ambiguity, trust, and the responsible conduct of science. *Science and Engineering Ethics* 5:205–214.

Korenman SG, Shipp A, eds. 1994. *Teaching the Responsible Conduct of Research Through a Case Study Approach: A Handbook for Instructors.* Washington, DC: Association of American Medical Colleges.

Kovac J. 1995. *The Ethical Chemist: Case Studies in Scientific Ethics.* Knoxville, TN: Department of Chemistry, University of Tennessee.

Macrina FL. 2000. *Scientific Integrity: An Introductory Text with Cases,* 2nd ed. Washington, DC: ASM Press.

National Academy of Sciences. 1995. *On Being a Scientist,* 2nd ed. Committee on the Conduct of Science. Washington, DC: National Academy Press. (Available for free download at the NAS website: http://www.nationalacademies.org/publications/.)

National Academy of Sciences, National Academy of Engineering, and Institute of Medicine. Panel on Scientific Responsibility and the Conduct of Research. 1992. *Responsible Science: Ensuring the Integrity of the Research Process,* Vol. I. Committee on Science, Engineering, and Public Policy. Washington, DC: National Academy Press.

National Academy of Sciences, National Academy of Engineering, and Institute of Medicine. 1993. *Responsible Science: Ensuring the Integrity of the Research Process,* Vol. II. Panel on Scientific Responsibility and the Conduct of Research. Committee on Science, Engineering, and Public Policy. Washington, DC: National Academy Press.

National Academy of Sciences, National Academy of Engineering, and Institute of Medicine. 2001. *Preserving Public Trust: Accreditation and Human Research Participant Protection Programs.* Committee on Assessing the System for Protecting Human Research Subjects. Washington, DC: National Academy Press.

Penslar RL, ed. 1995. *Research Ethics: Cases and Materials.* Indianapolis, IN: Indiana University Press.

Resnik DB. 1998. *The Ethics of Science: An Introduction.* New York: Routledge.

Sigma Xi. 1991. *Honor in Science.* Research Triangle Park, NC: Sigma Xi, The Scientific Research Society.
Sigma Xi. 1999. *The Responsible Researcher: Paths and Pitfalls.* Research Triangle Park, NC: Sigma Xi, The Scientific Research Society.

Teaching

Banner JM, Jr., Cannon HC. 1997. *The Elements of Teaching.* New Haven, CT: Yale University Press.
Bransford JD, Brown AL, Cocking RR. 1999. *How People Learn: Brain, Mind, Experience, and School.* Committee on Developments in Science of Learning and Commission on Behavioral and Social Sciences and Education. Washington, DC: National Academy Press.
McKeachie WJ, Gibbs G. 1998. *Teaching Tips: Strategies, Research, and Theory for College and University Teachers,* 10th ed. New York, NY: Houghton Mifflin.
Palmer PJ. 1998. *The Courage to Teach: Exploring the Inner Landscape of a Teacher's Life.* San Francisco, CA: Jossey-Bass.
Pregent R. 1994. *Charting Your Course: How to Prepare to Teach More Effectively.* Madison, WI: Magna Publications.
Shor I, Friere P. 1987. *A Pedagogy for Liberation. Dialogues on Transforming Education.* South Hadley, MA: Bergen & Garvey Publishers, Inc.
Stigler JW, Hiebert J. 2000. *The Teaching Gap: Best Ideas from the World's Teachers for Improving Education in the Classroom.* New York, NY: The Free Press.
Taylor PC, Gilmer PJ, Tobin K, eds. 2002. *Transforming Undergraduate Science Teaching: Social Constructivist Perspectives.* New York, NY: Peter Lang Publishing, Inc.

Writing

Booth V. 1993. *Communicating in Science: Writing a Scientific Paper and Speaking at Scientific Meetings,* 2nd ed. New York, NY: Cambridge University Press.
Council of Biology Editors, Committee on Graduate Education in Scientific Writing. 1968. *Scientific Writing for Graduate Students: A Manual on the Teaching of Scientific Writing.* New York, NY: Rockefeller University Press.
Day RA. 1998. *How to Write and Publish a Scientific Paper,* 5th ed. Phoenix, AZ: Oryx Press.

E

Committee and Staff Biographies

COMMITTEE BIOGRAPHIES

Arthur Rubenstein, M.B.B.Ch. (*Chair*), is executive vice president of the University of Pennsylvania for the Health System and dean of the School of Medicine, University of Pennsylvania School of Medicine. Previously, he was dean and chief executive officer of Mount Sinai School of Medicine and the Gustave L. Levy Distinguished Professor and served as executive vice president of Mount Sinai/New York University School of Health. He has also been chair of the Department of Medicine at the University of Chicago Pritzker School of Medicine and the Lowell T. Coggeshall Distinguished Service Professor of Medicine. He is an authority on diabetes, a widely sought counselor to academic health centers, and a frequent panelist at the annual meetings of the senior research societies in internal medicine. Dr. Rubenstein collaborated with Donald Steiner, who discovered proinsulin. He has extensively studied C peptide as a measure of endogenous insulin secretion. He has discovered several families with mutations in their insulin genes and has characterized these abnormalities. The widely used assay for the C peptide of insulin, developed in his laboratory, has provided a means of studying insulin metabolism in diabetic patients receiving exogenous insulin. For his research efforts, he has received numerous awards and named lectureships. Among these are the Eli Lilly Award and the Banting Medal of the American Diabetes Association and the David Rumbough Award of the Juvenile Diabetes Foundation. Dr. Rubenstein was elected a master of the American College of

Physicians in 1987 and received its John Phillips Memorial Award in 1995. He is a fellow of the College of Medicine of South Africa (1964), the Royal College of Physicians of London (1977), and the American Association for the Advancement of Science. He is past president of the Association of Professors of Medicine (1995–1996), from which he also received the Robert H. Williams Distinguished Chair of Medicine Award (1997); the Association of American Physicians (1995–1996); the Central Society for Clinical Research (1989); and the Chicago Society of Internal Medicine (1992–1993). He was a member of the National Institutes of Health (National Institute of Diabetes and Digestive and Kidney Diseases) Advisory Board, the National Institutes of Health Metabolism Study Section, and the National Diabetes Advisory Board and is a member of the Institute of Medicine, the American Academy of Arts and Sciences, and the Residency Review Committee in Internal Medicine. He served on the American Board of Internal Medicine for 8 years and was its chair in 1992–1993. He has authored more that 350 papers and has been on the editorial boards of *Annals of Internal Medicine, Journal of Diabetes and its Complications,* and *Medicine.*

Muriel Bebeau, Ph.D., is a professor in the Department of Preventive Sciences, School of Dentistry; executive director of the University's Center for the Study of Ethical Development, and a faculty associate in the Center for Bioethics. She received her undergraduate degree from Concordia College, River Forest, Illinois, and spent her early career as a musician and music educator. Her graduate degrees are from Arizona State University, where she held a faculty appointment before coming to the University of Minnesota in 1979. As an educational psychologist, Dr. Bebeau pioneered the teaching of ethics in dentistry. To evaluate outcomes, she and Jim Rest designed and validated measures that assess functional processes that give rise to morality. Recently, Dr. Bebeau and colleagues at Indiana University's Poynter Center applied ideas worked out in dentistry to research ethics. As chair of the American Association for Dental Research (AADR) Ethics Committee, she organized a symposium to explore the role of scientific societies in deterring misconduct and developed a consensus statement on the future directions of AADR to promote integrity in research. In recognition of her contributions to dental ethics, the American College of Dentists awarded her an honorary fellowship. The Association for Moral Education recognized her contributions to moral psychology with its lifetime achievement award. Dr. Bebeau's primary interests are studying the processes involved in ethical decision making (sensitivity, reasoning and judgment, commitment, and actions) and their roles as determinants of ethical behavior.

Stuart Bondurant, M.D., is professor of medicine and dean emeritus at the School of Medicine of the University of North Carolina at Chapel Hill (UNC-CH). He was a member of the faculty of the School of Medicine at Indiana University Medical Center and was chief of the Medical Branch of the Artificial Heart-Myocardial Infarction Program at the National Heart Institute. There he directed the establishment of the first national program of research on myocardial infarction. Dr. Bondurant was professor and chair of the Department of Medicine before serving as president and dean of Albany Medical College in Albany, New York. In 1979 he became professor of medicine and dean of the School of Medicine of UNC-CH. In July 1994 he completed three terms as dean and, while on leave of absence from UNC-CH, served as director of the Center for Urban Epidemiologic Studies of the New York Academy of Medicine. In 1996 and 1997, Dr. Bondurant served as interim dean of the UNC-CH School of Medicine. During his career he has served as an officer of many organizations and societies including president of the American College of Physicians, the Association of American Physicians, and the American Clinical and Climatological Association; acting president of the Institute of Medicine of the National Academy of Sciences; vice president of the American Heart Association and of the American Society for Clinical Investigation; chair of the board of the North Carolina Biotechnology Center; chair of the Council of Deans of the Association of American Medical Colleges; and chair of the Association of American Medical Colleges. From 1989 to 1995 he served as chair of the North Carolina Governors Commission on the Reduction of Infant Mortality, and from 1989 to the present he has served as vice chair of The Healthy Start Foundation. He also has served as adviser to the National Institutes of Health, the Veterans Administration (the U.S. Department of Veterans Affairs), the U.S. Department of Defense, and the U.S. Department of Health and Human Services. He is a master of the American College of Physicians and is a fellow of the Royal College of Physicians of Edinburgh and of the Royal College of Physicians of London. He holds an honorary doctor of science degree from Indiana University, the Citizen Laureate Award of the Albany (New York) Foundation, and the 1998 Thomas Jefferson Award of the Faculty of the University of North Carolina. Dr. Bondurant is a member of the Institute of Medicine of the National Academy of Sciences, and he received the David P. Rall Award from that organization in 2000.

David Cox, M.D., Ph.D., is scientific director of Perlegen Sciences. He is on a leave of absence from Stanford University, where he is professor of genetics and pediatrics at the Stanford University School of Medicine as well as codirector of the Stanford Human Genome Center. After receiving

A.B. and M.S. degrees from Brown University in Rhode Island, Dr. Cox obtained M.D. and Ph.D. degrees from the University of Washington, Seattle. Dr. Cox completed a pediatric residency at the Yale-New Haven Hospital in New Haven, Connecticut, and was a fellow in both genetics and pediatrics at the University of California, San Francisco. Dr. Cox is certified by both the American Board of Pediatrics and the American Board of Medical Genetics. Dr. Cox is an active participant in the large-scale mapping and sequencing efforts of the Human Genome Project while carrying out research involving the molecular basis of human genetic disease.

Robert C. Dynes, Ph.D., is professor of physics, former chair of the Physics Department, and chancellor at the University of California, San Diego. Before coming to the university in 1991, Dr. Dynes had a distinguished career at AT&T Bell Laboratories, where he held a variety of positions culminating in a 7-year term as director of Chemical Physics Research. His research on the properties of conductors and superconductors has led to more than 200 invited talks at national and international meetings, 175 publications in peer-reviewed journals, and seven patents. Professor Dynes is on the editorial boards of four physics journals and serves on several committees of national importance: The [University of California] President's Council on the National Laboratories, the Sloan Foundation Review Panel, the Department of Energy Council on Materials, the Advisory Board for the University of Texas Center for Superconductivity, and the Los Alamos National Laboratory Physics Division External Review Committee. A member of the National Academy of Sciences since 1989, Dr. Dynes did his undergraduate studies at the University of Western Ontario and earned both a master of science degree and a Ph.D. from McMaster University.

Mark S. Frankel, Ph.D., has been director of the Scientific Freedom, Responsibility and Law Program at the American Association for the Advancement of Science (AAAS) since 1990, where he develops and manages the association's activities related to professional ethics, science and society, and science and law. He is a AAAS fellow, editor of the association's quarterly publication, *Professional Ethics Report*, and staff officer for two AAAS committees: the Committee on Scientific Freedom and Responsibility and the AAAS-American Bar Association National Conference of Lawyers and Scientists. He is on the board of directors of the National Patient Safety Foundation and serves on the editorial boards of *Professional Ethics, Ethics and Behavior, Science and Engineering Ethics,* and *Law and Human Genome Review.* Dr. Frankel has directed several AAAS projects and has published extensively on integrity in research and scien-

188

tific misconduct. The AAAS videos "Integrity in Scientific Research" are widely used throughout the United States and abroad as part of educational programs on research ethics.

Penny J. Gilmer, Ph.D., is professor of chemistry and biochemistry at Florida State University in Tallahassee. She received a B.A. in chemistry at Douglass College, an M.A. in organic chemistry at Bryn Mawr College, and a Ph.D. in biochemistry at the University of California, Berkeley. Her research interests include cell-surface biochemistry, science education, and ethics in science. Since 1984 she has taught both undergraduates and graduate students on ethical issues in science, including topics such as the development of the atomic bomb, issues in human subjects and animals in research, and environmental ethics. She serves on the editorial board of *Science* and *Engineering Ethics*. Dr. Gilmer is finishing the writing for a second doctorate in science education through Curtin University of Technology in Australia. She has studied her own World Wide Web-enhanced biochemistry classroom using qualitative research methodology. In 1999 Professor Gilmer received the Innovative Excellence in Teaching, Learning and Technology Award from the 10th International Conference on College Teaching and Learning. She coedited the recently published book *Transforming Undergraduate Science Teaching: Social Constructivist Perspectives* (Peter Lang, 2001). Dr. Gilmer also is co-principal investigator of the National Science Foundation-funded grant entitled the Florida Collaborative for Excellence in Teacher Preparation, whose goal is to increase the number and quality of teachers of mathematics and science for secondary schools in Florida.

Frederick Grinnell, Ph.D., is professor of cell biology and director of the Program in Ethics in Science and Medicine at The University of Texas Southwestern Medical Center in Dallas, where he has worked since earning a doctorate in biochemistry from Tufts University in 1970. Research in his laboratory concerns the wound-repair process, which has resulted in three patents, a MERIT award from the National Institute of General Medical Sciences, and more than 100 publications in peer-reviewed journals. Dr. Grinnell's ethics-related research program attempts to articulate what doing science entails, with the goal of informing science policy decisions and advancing science education and public understanding of science. His book *The Scientific Attitude* (Guilford Press, 1992) is now in its second edition. A sequel, *Everyday Practice of Science*, is in preparation. He has taught a graduate course on the philosophy and conduct of science for more than 20 years and has participated in teaching medical ethics to medical students for a decade. He served on the university's institutional review board as a member and alternate for 10 years and is a member of its Review Committee on Conflict of Interest. Nationally, Professor Grinnell is a member of the Science Policy Committee of the Federation of American Societies of Experimental Biology (FASEB) and previous chair of the FASEB Subcommittee on Human Subjects and Bioethics. He is also a member of the National Institutes of Health Peer Review Oversight Group. Dr. Grinnell drafted the Code of Ethics adopted by the American Society of Biochemistry and Molecular Biology in 1998.

Joyce Miller Iutcovich, Ph.D., is a sociologist who followed a dual career path for many years as a university professor as well as president of the Keystone University Research Corporation (KURC), an independent research and consulting organization. As an academic, Dr. Iutcovich achieved full professor rank at Gannon University in 1991 and later served as associate provost from 1995 to 1997. In 1999 Dr. Iutcovich made the decision to leave the academy on a full-time basis to dedicate her time to KURC, with its ever-expanding operations. As president of KURC, Dr. Iutcovich has presided over the company's growth from modest beginnings to annual revenues of more than $10 million. Projects undertaken by KURC include short-term applied social research and consulting activities, as well as multiyear contracts to design and administer programs for state government agencies. In her capacity as an applied social researcher, Dr. Iutcovich has expertise in evaluation research and social policy, primarily in the social service and educational fields. She is active in the sociological professional associations and served as president of the Society for Applied Sociology from 1994 to 1995. In 2002 she was awarded a congressional fellowship from the American Sociological Association and worked with Senator Jack Reed (D-RI) for the term of her fellowship. Her professional writings have appeared in such journals as *Evaluation and Program Planning, Journal of Applied Gerontology, Journal of Applied Sociology, Sociological Practice, Handbook of Clinical Sociology,* and *Social Insight: Knowledge at Work.* She coauthored *The Sociologist as Consultant* (Praeger, 1987) and coedited *Directions in Applied Sociology* (Society for Applied Sociology, 1997). She was a member of the Committee on Professional Ethics (COPE) of the American Sociological Association from 1993 to 1999 during an extensive code revision process; she served as the chair of COPE in 1999.

Stanley G. Korenman, M.D., is professor of medicine and associate dean for ethics and the Medical Scientist Training Program at the University of California, Los Angeles (UCLA). He is also the founding chair of the Ethics Advisory Committee of the Endocrine Society. Board certified in internal medicine with an endocrinology subspecialty, he has published more than 100 peer-reviewed papers in basic and clinical reproductive endocrinology since receiving an M.D. from Columbia University. Dr. Korenman previously served as chief of the Department of Medicine of the UCLA San Fernando Valley Program and chief of the Medical Service at the Veterans Administration Medical Center in Sepulveda, California; as chief of endocrinology at UCLA and the University of Iowa; as director of the Harbor-UCLA Clinical Research Center; and as a senior investigator at the National Cancer Institute. Over the past few years he has published a number of empirical and philosophical articles on scientific ethics and misconduct as well as a widely used text on teaching the responsible conduct of research.

Joseph B. Martin, M.D., Ph.D., was born in Bassano, Alberta, Canada, in 1938. Dr. Martin received his premedical and medical education at the University of Alberta, Edmonton, receiving the M.D. degree in 1962. He completed a residency in neurology in 1966 and a fellowship in neuropathology in 1967 at Case Western Reserve University in Cleveland, Ohio, and earned a Ph.D. in anatomy from the University of Rochester in 1971. Dr. Martin began his distinguished career in academic medicine at McGill University in Montreal, Quebec, Canada, where he eventually became chair of the Department of Neurology and Neurosurgery in 1977. In 1978, he joined the faculty of Harvard Medical School in Boston, where he held the title first of Bullard Professor of Neurology and chief of the Neurology Service at Massachusetts General Hospital. In 1984, Dr. Martin was appointed the Julieanne Dorn Professor of Neurology at Harvard. He was appointed dean of the Harvard Faculty of Medicine effective July 1, 1997. Before returning to the Harvard medical community, he served as chancellor of the University of California, San Francisco (UCSF), for 4 years. Dr. Martin initially went to UCSF in 1989 as dean of the School of Medicine. He is a member of the Institute of Medicine and a fellow of the American Academy of Arts and Sciences and the American Association for the Advancement of Science.

Robert R. Rich, M.D., is executive associate dean and professor of medicine and of microbiology and immunology at Emory University School of Medicine. He received an M.D. from the University of Kansas and completed an internship and residency in internal medicine at the University of Washington. He had subspecialty training in allergy and immunology and postdoctoral fellowships at the National Institutes of Health (NIH) and Harvard Medical School. In 1973, Dr. Rich joined the faculty of Baylor College of Medicine as assistant professor of microbiology and immunology and of medicine. In 1978 he was promoted to professor and in 1995 was named Distinguished Service Professor. He was an investigator of the Howard Hughes Medical Institute from 1977 to 1991. From 1990 to

1998 he also served as vice president and dean of research at Baylor. He is deputy editor of *The Journal of Immunology* and is its editor-in-chief designate. He has had appointments to other to editorial boards, including those of *The Journal of Experimental Medicine, The Journal of Infectious Diseases, The Journal of Clinical Immunology,* and *Clinical and Experimental Immunology.* He is also editor-in-chief of a comprehensive textbook, *Clinical Immunology: Principles and Practice* (Mosby-Year Book, 1996). He served as a member and chairman of two NIH study sections and on the boards of directors of the American Board of Allergy and Immunology (of which he was also chairman), the American Board of Internal Medicine, and the National Space Biomedical Research Institute. He is past president of the Clinical Immunology Society. Dr. Rich is a member of the Advisory Panel on Research of the Association of American Medical Colleges and is a member of the National Human Research Protections Advisory Committee of the U.S. Department of Health and Human Services. He is a councilor of the American Clinical and Climatological Association. He is vice president of the American Academy of Allergy, Asthma and Immunology and is past chairman of its Professional Education Council. He is also president of the Federation of American Societies for Experimental Biology, the largest coalition of biomedical research scientists in the United States.

Louis M. Sherwood, M.D., is senior vice president for medical and scientific affairs in the U.S. Human Health Division of Merck & Co. Having previously led new drug development as executive vice president of the Merck Research Laboratories (1989 to 1992), Dr. Sherwood serves as chief medical officer for Merck & Co. in the United States and is responsible for all medical activities of Merck & Co. in the United States, including clinical development, outcomes research and management, medical services, academic and professional affairs, and the medical directors in various regions of the country. He also serves as adjunct professor of medicine at the University of Pennsylvania and visiting professor of medicine at the Albert Einstein College of Medicine. Before joining Merck & Co. in 1987, he served for 7 years as the Ted and Florence Baumritter Professor and chair of the Department of Medicine, Albert Einstein College of Medicine, and as physician-in-chief, Montefiore Medical Center, New York, New York. He had previously served as chair of the Department of Medicine at the Michael Reese Medical Center and professor of medicine at the University of Chicago, as well as associate professor of medicine at Harvard Medical School and chief of the Endocrine Unit at Beth Israel Hospital. Dr. Sherwood has been president of the American Society for Clinical Investigation, the Association of Program Directors in Internal Medicine, and the American Physicians Fellowship for Medicine in Israel; and he also serves on the Clinical Research Roundtable of the Institute of Medicine and the National Research Advisory Council of the U.S. Department of Veterans Affairs. He has received numerous awards, the most recent of which are the Lifetime Achievement Award from the Academy of Pharmaceutical Physicians (2001) and the Special Recognition Award from the Association of Professors of Medicine (2002). Dr. Sherwood received an undergraduate degree from Johns Hopkins University and an M.D. from the College of Physicians and Surgeons of Columbia University. He has published extensively in the field of endocrinology, particularly on the regulation of parathyroid hormone synthesis and secretion and clinical calcium disorders. Dr. Sherwood retired from Merck & Co. on April 1, 2002 and is now working as an independent consultant.

Michael J. Zigmond, Ph.D., is a Professor of Neurology, Psychiatry, and Neurobiology at the University of Pittsburgh. His area of research interest is neuronal death and neuroprotection as it applies to neurodegenerative diseases, particularly Parkinson's disease. He has held grants from the National Institutes of Health (NIH) since 1975, including a Research Scientist Award (1985-2001), and currently holds a MERIT Award from NIMH. He is co-director of the National Parkinson Foundation Center of Excellence at the University of Pittsburgh, and associate director of the Pittsburgh Institute for Neurodegenerative Disorders. He has been on several scientific advisory boards including the Dystonia Medical Research Foundation, Tourette Syndrome Association, and the Michael J. Fox Foundation, a member of several review committees of the NIH, and has served as the secretary of the Society for Neuroscience. Dr. Zigmond also has been involved in a variety of educational activities. He was the president of the Association of Neuroscience Departments and Programs (1990-91) and the organizing editor for *Fundamental Neuroscience* (1999). He was a member of the external advisory committee to the Committee on Science, Engineering, and Public Policy (COSEPUP) when it produced its manual, *Enhancing the Postdoctoral Experience for Scientists and Engineers,* and has advised COSEPUP on other projects related to education. He currently directs several NIH-funded training grants for graduate students and postdoctoral fellows in the neurosciences at the University of Pittsburgh, and co-directs with Ms. Beth A. Fischer the university's Survival Skills and Ethics Program. In addition, he serves as the chair of the International Advisory Committee of the Society of Neuroscientists of Africa, has been active in training activities in Africa since 1993, and co-authors with Ms. Fischer a monthly electronic column on professional skills for the International Brain Research Organization. Dr. Zigmond is currently editor-in-chief of *Progress in Neurobiology* and sits on the editorial boards of a dozen other professional journals. Dr. Zigmond received his B.S. in chemical engineering from Carnegie Institute of Technology (now Carnegie Mellon University), his Ph.D. in biopsychology from the University of Chicago, and his postdoctoral training at the Massachusetts Institute of Technology. He was a Visiting Fellow in Neuroscience at the Children's Hospital in Boston, Massachusetts from 1998-1999. Dr. Zigmond's contributions in the area of professional development and responsible conduct and ethics include courses and publications on promoting responsible conduct of research and on scientific publishing and other professional skills. Dr. Zigmond chaired the committee of the Society for Neuroscience that developed the society's guidelines for responsible conduct in publishing and now chairs the society's Social Issues Committee.

Board on Health Sciences Policy Liaison

Ada Sue Hinshaw is a nationally recognized contributor to nursing research and is dean and professor at the University of Michigan School of Nursing. Before coming to the University of Michigan, Dr. Hinshaw was the first permanent director of the National Institute of Nursing Research at the National Institutes of Health in Bethesda, Maryland. Dr. Hinshaw led the institute in its support of valuable research and training in many areas of nursing science, such as disease prevention, health promotion, acute and chronic illness, and the environments that enhance nursing care patient outcomes. From 1975 to 1987, Dr. Hinshaw served as director of research and professor at the University of Arizona College of Nursing in Tucson and as director of nursing research at the University Medical Center's Department of Nursing. She has also held faculty positions at the University of California, San Francisco, and the University of Kansas. Dr. Hinshaw received a Ph.D. and M.A. in sociology from the University of Arizona, an M.S.N. from Yale University, and a B.S. from the University of Kansas. Her major fields of study included maternal-newborn health, clinical nursing and nursing administration, and instrument development and testing.

IOM STAFF

Theresa M. Wizemann, Ph.D., is a senior program officer for the Board on Health Sciences Policy at the Institute of Medicine (IOM) and served as the study director for *Integrity in Scientific Research*. Previously, she was the study director for the IOM study *Exploring the Biological Contributions to Human Health: Does Sex Matter?* Dr. Wizemann came to IOM from the office of Senator Edward M. Kennedy, Senate Committee on Health Education, Labor and Pensions, where she handled various health and science policy issues as a congressional fellow. Prior to the fellowship, she led a vaccine research team at MedImmune, Inc., a leading biotechnology company in Maryland. She earned a bachelor's degree in medical technology from Douglass College of Rutgers University and master's and doctoral degrees in microbiology and molecular genetics, jointly from Rutgers University and the University of Medicine and Dentistry of New Jersey. She did a postdoctoral fellowship in infectious diseases at The Rockefeller University in New York. Dr. Wizemann has expertise in microbiology, immunology, and infectious diseases and has a particular interest in women's and children's health.

Mehreen Butt is a senior project assistant for the Board on Health Sciences Policy. Before joining the Institute of Medicine she worked for an international nonprofit health organization that assisted developing countries in restructuring their health care delivery systems. She received bachelor of science degrees in Biology and English from Tufts University in Medford, Massachusetts.

Andrew Pope, Ph.D., is director of the Board on Health Sciences Policy at the Institute of Medicine. With expertise in physiology and biochemistry, his primary interests focus on environmental and occupational influences on human health. Dr. Pope's previous research activities focused on the neuroendocrine and reproductive effects of various environmental substances on food-producing animals. During his tenure at the National Academy of Sciences and since 1989 at the Institute of Medicine, Dr. Pope has directed numerous reports on topics that include injury control, disability prevention, biologic markers, neurotoxicology, indoor allergens, and the enhancement of environmental and occupational health content in medical and nursing school curricula. Most recently, Dr. Pope directed studies on priority-setting processes at the National Institutes of Health, fluid resuscitation practices in combat casualties, and organ procurement and transplantation.

Index

investigations and inspections, 35, 41,
45, 54, 73-74, 79, 168
licensing, 25, 37, 80, 118
occupational health and safety, 73, 89
open systems model, codes of conduct,
54-55, 58-59, 61, 66
open systems model, other, 53, 63, 64,
65
reporting requirements, 73-75
sanctions, 35, 41, 59-60, 67, 76
self-assessment and, 10-11, 73
Religious issues, 38
Reporting requirements, 5, 34, 36, 37, 79,
107, 171, 172
bibliographic resources on writing, 183
journals, 38, 63, 64, 66, 85, 101, 105
mentoring, 40
regulatory compliance, 73-75
*The Responsible Conduct of Research in the
Health Sciences*, 20, 167-177
*Responsible Science: Ensuring the Integrity of
the Research Process*, 19, 20-21, 22-
23, 140
Reviews of the literature, *see* Literature
reviews
Reward systems, 20-21, 24, 76, 112, 127, 175
Award for Excellence in Human
Research Protection, 77
Malcolm Baldrige Quality Award, 77-78
National Research Service Awards, 84

S

Sanctions, 35, 41, 59-60, 67, 76
The Scientific Attitude, 95
Scientific Research Society, 21
Self-assessment, 3, 10-11, 78, 79-82, 112-123
see also Peer review
accountability, general, 10, 73
administrators, 79, 116, 121
economic factors, general, 80-81, 113
funding, 73, 118-120, 122
institutions, 10, 11, 13, 73, 78, 112, 113-
122, 124, 130-132
professional education, 79, 80-81, 113-
120
regulatory issues, 10-11, 73
standards, 79, 80
technology, 79
Self-efficacy abilities, 96-97

Self-regulation abilities, 96-97
Sex differences, *see* Gender factors
Sigma Xi, *see* Scientific Research Society
Social factors, *see* Cultural issues;
Institutional factors; Mentor/
trainee relations; Political factors;
Public opinion
Society for Neuroscience, 106
Standards, 24, 33, 34, 58, 73, 81, 84, 115,
169-170, 174-175, 176
see also Accreditation, institutional;
Outcome measures; Performance-
based assessment
benchmarks, 75-77, 78-79, 81, 112
codes of conduct, 54-55, 58-59, 61, 66,
95, 120, 127, 138
identity formation, 94
licensing, 25, 37, 80, 118
open systems model, codes of conduct,
54-55, 58-59, 61, 66
self-assessment, 79, 80

T

Technology
computer-based instruction, 103-104,
151
computer technology, general, 25
inadequate, 17
Internet, 21, 105, 125, 138, 170
open systems model, 54, 57, 67
public opinion, 1, 16
self-assessment, 79
Textbooks, 106
Theoretical issues, 3, 6, 9, 49, 68, 98, 103,
125
see also Models and modeling

U

Universities and colleges, 25, 27, 39-40, 81-
82, 84-111, 116-117, 120-122, 148-
149
committee biographies, 184-194
conflicts of interest, 38, 43-44
open systems model, 50, 55
regulatory compliance, 74
U.S. Army, 149-150

V

Varmus, Harold, 17
Videotapes, 90, 151-152, 188

W

Wellcome Trust, 21
Women, *see* Gender factors
Workshops, 61, 97, 103, 105, 106
World Wide Web, *see* Internet